PROJECT EXECUTION

PROJECT EXECUTION
A PRACTICAL APPROACH TO INDUSTRIAL
AND COMMERCIAL PROJECT MANAGEMENT

Dr. Chitram Lutchman
DBA, MBA, B.Sc., CSP, CRSP, 1st Class Power Engineer

CRC Press
Taylor & Francis Group
Boca Raton   London   New York

CRC Press is an imprint of the
Taylor & Francis Group, an **informa** business

CRC Press
Taylor & Francis Group
6000 Broken Sound Parkway NW, Suite 300
Boca Raton, FL 33487-2742

© 2011 by Taylor and Francis Group, LLC
CRC Press is an imprint of Taylor & Francis Group, an Informa business

No claim to original U.S. Government works

Printed in the United States of America on acid-free paper
10 9 8 7 6 5 4 3 2 1

International Standard Book Number: 978-1-4398-3863-1 (Hardback)

Library of Congress Cataloging-in-Publication Data

Lutchman, Chitram.
　　Project execution : a practical approach to industrial and commercial project management / Chitram Lutchman.
　　　　p. cm.
　　Includes bibliographical references and index.
　　ISBN 978-1-4398-3863-1
　　1. Project management. I. Title.

HD69.P75L88 2011
658.4'04--dc22
　　　　　　　　　　　　　　　　　　　　　　　　　　　　　　　2010004471

Visit the Taylor & Francis Web site at
http://www.taylorandfrancis.com

and the CRC Press Web site at
http://www.crcpress.com

This book is dedicated to my family and my parents and the many individuals who have encouraged my learning. To my wife Sita, for her steadfast support and companionship. To my children Kevan, Megan and Alexa for their patience. To my friends and colleagues whose experiences and insights also added to the knowledge provided in this book. To the many hardworking project management personnel who strive tirelessly to deliver projects safely and on budget and on schedule.

Executive Summary

This book focuses on the essential requirements for successful execution of commercial and industrial projects. This book identifies people, process, and system readiness as key components of an overall milestone readiness process which, when managed properly, greatly increases the ability of project leaders to deliver projects *on budgets and on schedules*. The readiness process applies to the project execution stages of any commercial or industrial project regardless of industry. This book highlights practical measures and tools that can be used by project leaders to promote smooth and controlled execution of a project such that all stakeholders can be rewarded through a collective value maximization effort. Tools and measures discussed in this book are both scalable and transferable across industries. This book also identifies strong leadership behaviors and stakeholder relations and management as key requirements for successful project execution. The recommendations provided in this book are based on the practical experiences of the author and are intended to prompt project leaders to consider project execution requirements that are generally intuitive, but very likely to be forgotten given the many competing priorities of project leaders. This book provides a great reminder for seasoned project leaders and an excellent learning opportunity for inexperienced and junior project professionals to enhance their preparedness, skills, and capabilities in project execution roles. Finally, this book is a great reminder to personnel such as operators and maintenance staff involved in supporting the execution stage of a project.

Contents

List of Figures

List of Tables

Acknowledgments

I humbly acknowledge the help, support, compromises, and encouragement of my wife Sita and children Kevan, Megan and Alexa in allowing me to complete this book. My sincere thanks to Don Clague, vice president, in situ operations, Suncor Energy Inc., for his many direct and indirect contributions to improving this book. My sincere thanks to Carl Lussier and Reidar Hustoft for reviewing and contributing to the contents of this book. Sincere thanks to Dr. Beth Reaves, Andy Britton, Motilal Kissoon, and Peter Smalley for their editorial contributions and improvement suggestions. Thanks also to Sue MacKenzie, former vice president, human resources, Petro-Canada and former project leader, for an excellent project learning opportunity. Sincere thanks also to Ted Doleman, Glen Walker, and Kirk Taylor for the great learning opportunities they provided by supporting me in a leadership role in a project environment.

I would also like to acknowledge the contributions of Petro-Canada in allowing me to use some tools and concepts developed within the organization. These tools and concepts have added great value to this book. My thanks also to Ash Sharma for his efforts in providing pictures which added to the illustrations provided in this book. Finally, my sincere thanks to my many coworkers who have shared their experiences and knowledge with me in the project world and have inspired my learning.

Introduction

Having worked in project management roles for more than 20 years, I have found this area of the business to be both exciting and challenging. I have found also that project management provides an excellent avenue for improving stakeholder confidence in an organization. From my experience in various roles in project management and across different industries, managing stakeholder interest and expectations is critical to the overall success of any project. The ability to deliver projects on schedule and on budget has a profound effect on stakeholder confidence. Successfully managed projects often generate confidence in the market, leading to a favorable organizational image and general growth of the organization. The yardstick by which success is measured is essentially the same across industries and is often determined by cost control and scheduled delivery.

There are many models and tools available for managing projects in the industrial world today, all of which have been used with varying levels of success across their relevant sector. In spite of the many choices, however, cost overruns and delayed schedules are not uncommon. This is particularly true in the energy industry. One needs only to look at the oil sands of Alberta to understand the magnitude of this problem. Large industrial organizations like Syncrude Canada, Albian Sands, Canadian Natural Resources Limited (CNRL), and the Opti/Nexen Longlake projects have all experienced significant capital cost overruns on development projects. The Alberta Economic Development Authority (2004) pointed to "20–100% cost/schedule overruns" (p. 36) for large development projects across Canada.

While today's current economic environment has dampened investments in some regions, the widening gap between demand and supply of some energy commodities will lead to a resurgence of investments in energy-rich regions. Consider the oil sands of northern Alberta, which hold an untapped reserve of 1.7 trillion barrels of heavy oils. Up to mid-2008, the region was abuzz with commercial and industrial activities. According to Carter (2007), capital cost projections for "oil sands projects from 2006 to 2015 will total $110 billion" (p. 69). Should lessons learned from previous projects in this regional environment fail to infiltrate employers, we can expect this projection to increase by tens of billions of dollars before these ambitious projects begin to generate stakeholder value.

This book is designed to provide a practical approach for managing large industrial projects and focuses primarily on the execution (implementation-and-control) stage of the project cycle, since this is the stage of project management during which cost overruns and schedule delays often occur. The content of this book draws on my field experience in project management and uses proven principles for early identification of potential derailment variables so that they may be addressed before cost and schedule impacts are felt. This book provides a proven model for practical project execution. The recommendations and tools provided help to improve a project leader's ability to deliver projects within budget and on schedule even in hostile business environments. The principles of the book focus on early identification and correction of potential issues or concerns that can affect cost and schedule of the project.

This book differs from other project management books by focusing exclusively on the execution stage of the project as an excellent opportunity for value maximization for the organization. The execution stage of a project sees the transfer of a concept, idea, or process into physical structures and tangible entities. Land is cleared, buildings are constructed, equipment and machinery are constructed and installed, and a production process of some sort begins.

All of these activities result in money being spent. When construction equipment and personnel are mobilized, the goal of any good project leader is to have them complete their jobs in the most effective and efficient manner possible and to see them leave in the shortest possible time. However, there is no benefit in having the construction organization build the facility or complete the project without someone to operate it. While construction is taking place, therefore, a project leader must simultaneously ensure the facility is properly staffed just in time for start-up of the facility. This means that the project leader must be able to hire, train, and ensure competency of adequate numbers of trained and qualified personnel to operate the facility at the time construction of the facility is completed.

A final ingredient is required beyond the completed construction of the facility and the availability of trained and competent people to operate the facility or project. The facility must be equipped with relevant processes to sustain production and output from the facility. For the project leader, this means that while the facility is being constructed and people are being hired and trained, the supporting operating tools or business processes must also be developed such that they are available for use at the same time the construction of the facility has been completed. In essence, while the three activities of construction (systems if the facility is being built), hiring and training (people), and operating tools (processes) may begin at different times during the execution stage, they must all be completed at the same time.

Achieving this critical goal is by no means an easy task. Even if this illustrious goal is achieved, the project leader can only boast of an on-schedule delivery of the project. The other half of the boast lies in the ability to manage cost throughout the process to deliver the project on budget and maximize stakeholder value from the project management process.

Managing the budget is a twofold process in the project execution stage. First, cost management is an absolute necessity. Hence, procurement practices, contractor and service provider management, management of human performance, and possession of the right tools to support the cost management process are absolutely essential. Second, a strong focus on the project schedule is required in managing the budget. A large element of the cost structure resides in the fixed cost associated with the mobilization of construction equipment and infrastructure. The construction industry is not in the business of subsidizing owners and clients. When schedules are delayed because of owner-related causes, fixed cost associated with equipment and machinery mobilized by the construction organization are passed on to the owner. Such cost can be exceptionally large. A drilling rig in the oil and gas industry, for example, can easily cost an owner $100,000 per day when idle with zero value-added contributions.

In addition to the fixed cost associated with equipment and machinery, there are costs associated with construction labor productivity. With each day of delay incurred, the cost of the construction workforce is generally passed on to the owner unless the root cause of the delay is attributable to the construction organization. In addition, when an owner is not able to move a project forward because of owner-related reasons, the cost of the construction labor force must be borne by the owner until the project can be moved forward. It is imperative, therefore, that owner-related delays are minimized if not avoided completely.

In a competitive market, an owner may be forced to absorb this cost because skilled construction trade workers may not be available to continue working on the project if they are temporarily laid off. More important to the cost management process, it must be noted that skilled construction trade workers are highly paid with wages ranging from $25 an hour to as high as $150 an hour. At these rates, a project leader cannot afford to have people idle during the execution stage. Moreover, during peak periods in the execution stage, the numbers of construction personnel on a project site can amount to thousands depending on the scale of the project.

To achieve the critical on-budget and on-schedule goal, therefore, project leaders must be equipped with the right tools, skills, and knowledge to deliver on all three components at the same time. Perhaps one of the most difficult elements of cost control during the execution stage is the ability of project leaders to maximize human performance. Leaders must be skilled in leading a large and culturally diverse workforce. They must be adequately trained to motivate workers to high levels of performance while being prescriptive and occasionally dictatorial in their leadership approaches. Conventional leadership theory advises that the present-day workforce has transcended the boundaries of autocratic leadership styles ("tell me") through the era of early engagement ("show me") to the present needs for involvement ("engage me") of the workforce.

The current leadership environment therefore suggests that the workforce expectations from leaders are higher. To meet these higher demands, the efforts required from leaders to meet the workforce expectations are also higher. As a consequence, the new breed of project leaders must be skilled leaders with exceptional technical capabilities to be able to deliver projects on schedule and on budget.

This book provides a compendium of tools and tips for the project leader to successfully meet the challenges of the execution stage. Practical methods, tools, and tested principles are provided to guide and support the performance of project leaders during the execution stage of a project.

This book also presents a tested alternative to the traditional Gantt charts and PERT (project evaluation review technique) charts for ensuring scheduled delivery of a project. This book offers the *readiness model* for supporting scheduled delivery of a project. The model focuses on ensuring readiness for people, processes, and systems (PP&S) at successive milestone events during the execution stage for ensuring scheduled delivery of a project. This model allows the project leadership team early identification of project issues in the areas of people, processes, or systems that can potentially lead to delays and cost overruns should they go unaddressed. In this way, resource allocation and focus can be brought to bear on areas that require priority actions so that ultimately projects are delivered on budget and on schedule. The readiness model allows the project team to perform readiness reviews on PP&S at specific and clearly defined points in the execution stage to allow adequate opportunities to realign resources in a timely fashion, thereby facilitating appropriate corrective and remedial actions as and when required.

The exciting thing about this readiness model is that it is practical, adaptable to different industries, and scalable depending on the size and scope of a project. This model has been proven to deliver success on large and small commercial projects. Furthermore, the model caters to the regulatory and stakeholder needs of the prevailing and future business environments. However, by far the most exciting attribute of this model is that it presents a viable option to project managers and leaders for managing cost overruns during the execution stages of projects. On-budget, on-schedule delivery in project management is achievable with the practical tools, ideas, and concepts provided in this book.

In this book, I share my experiences from managing many project workforces; these experiences provided the introduction to some useful tools adaptable to any industry or project environment. These tools are simple and practical to use. I examine the leadership environment in which we now operate and the changing leadership needs of the workforce. I also assimilate practical people and process learning and knowledge that can help project leaders meet their project goals of on-budget and on-schedule delivery. Included in this discussion is an explanation of how the readiness process works and the simple tools required to measure and communicate people readiness, process readiness, and system readiness for preferential resource allocation and course corrections as required.

The book also includes, in the project leader's tool kit, an acknowledgment of the need to ensure proper stakeholder management. Failure to manage stakeholders and stakeholders' expectations can result in severe undesirable consequences. The importance of stakeholder management cannot be understated and is a critical component of a project leader's ability. Failure to address stakeholder interests can result in project shutdowns, cost overruns, schedule delays, and overall poor project execution performance.

Finally, the book proposes a situ-transformational approach to leadership in the project execution environment. This leadership approach includes core concepts of Ken Blanchard's situational leadership model with transformational leadership behaviors. Key to success is to know the maturity level of each worker and to apply the right sets of transformational leadership behaviors to generate the best performance from workers during project execution.

About the Author

Dr. Chitram (Chit) Lutchman is an experienced project management professional in the oil and gas and agricultural industries. With more than 20 years experience leading and working with personnel at the front line, he has been involved in many macro- and microlevel projects.

Dr. Lutchman began his career in the oil and gas industry in Trinidad and Tobago, where he gained his first exposure to project execution principles and challenges. Subsequently, he left the oil and gas industry to assist in diversifying Trinidad's sugar industry, pursuing an import substitution strategy for replacing beef imports into the country.

After migrating to Canada in 1995, Dr. Lutchman reentered the oil and gas industry, becoming more involved in megaproject execution and leading the operations team in the commissioning and startup of a large commercial steam-assisted gravity drainage (SAGD) project in the oil sands of Alberta. He also led the integration of a cogeneration project into this operating SAGD facility. He subsequently became actively involved in corporate business planning and project monitoring.

Dr. Lutchman continues to work in other areas of the oil and gas industry. Some of his more recent ventures involved leading corporate safety and emergency management for Petro-Canada with responsibilities for enhancing the safety culture of the organization. At present, he is working on developing a system for enhancing contractor safety management for Suncor Energy Incorporated. Contractor management has been identified as an area of extreme risk exposure that can influence project performance during project execution.

Dr. Lutchman holds a bachelor of science degree in agricultural sciences, a first-class power engineer license, Canadian Registered Safety Professional (CRSP) and Certified Safety Professional (CSP) designations, a master of business administration degree from McMaster University, and a doctoral degree in business administration from the University of Phoenix.

1

The Project Cycle

Introduction

New and growing organizations are generally well rewarded by the market in the form of higher stakeholder confidence and share prices when projects are delivered on budget and on schedule. Consequently, organizations place huge emphasis on getting it right the first time and at the publicly disclosed price for completion of projects. Failure to get it right the first time results in rework, cost increases, schedule changes, and severe penalties from shareholders and other stakeholders. For the large organization with deep pockets, much of the cost increase in the short term can be funded from its cash flows and is eventually passed on to the shareholders in the form of reduced earnings and smaller shareholder value.

Smaller, more vulnerable organizations are more harshly judged on their project management expertise and capabilities. Often, such organizations are subjected to severe consequences resulting from cost overruns and delayed projects. It is not uncommon to see these smaller organizations go bankrupt, lose significant market share and value, be gobbled up by other industry players, or experience leadership changes within the organization in response to escalating cost and delays.

To the owners and shareholders of private companies, cost overruns and delays in project schedule often represent lost income, deferred earnings, and declining value. It is in our collective best interest, therefore, to continually hold management to higher levels of accountability in project management since poor performance in large-scale projects can translate into significant losses to the shareholder and reduced value to the organization.

We are all aware that the market responds to good news. We often see stock value and share price leap at the announcement of good news. On budget and on schedule are powerful announcements to the market to promote and fuel sustained growth of an organization. This must be the goal of every organization when project work is undertaken. To do so, however, requires excellent project leadership skills, a committed and focused workforce, a supporting business environment, a vision that is shared by all stakeholders, and often long, exhausting work days over extended periods.

1

Practical application of this readiness model has been demonstrated in the oil and gas sector. Nevertheless, the model is adaptable to most industries that may consider undertaking capital expansion projects, new technology adoption, upgrades, joint operations, and greenfield projects. Discussion here focuses primarily on the execution (implementation-and-control) stage of the project cycle; I provide examples of simple tools that any organization can develop in house to assess readiness on an ongoing basis so that project management can become a more simplistic task with more acceptable results for both the organization and its stakeholders.

The Simplified Project Cycle

The project cycle essentially regards the project environment as continuous. Four major stages in the cycle are planning, scoping and design, execution (implementation and control), and closeout (evaluation and feedback). A precursor to the planning stage is an identification phase. Identification is often included as an element of the planning stage but may also reside outside the planning stage. Figure 1.1 provides an overview of a simple project cycle. Figure 1.1 also shows some of the key activities that are performed

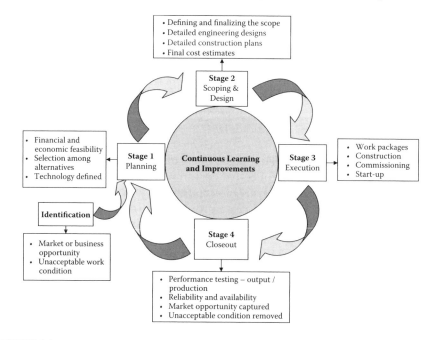

FIGURE 1.1
A simplified project cycle.

during each stage of the project cycle. As pointed out, planning is usually the first stage of the project cycle and generally results from the identification of either an undesirable condition or the existence of a business or market opportunity.

Identification and Planning

When a project idea is developed in an organization, it is often in response to an existing deficiency or a desire to capture new or existing market benefits and advantages. During this stage of the project cycle, business leaders are busy scanning the business environment, seeking opportunities to improve the overall performance and position of the organization.

The business environment today requires leaders to be proactive and vigilant. Opportunities can be lost quite easily by leaders who fail to grasp opportunities when they become available. In the identification and planning stages of the project cycle, seed ideas are rendered into potential value for consideration by senior leadership. Business planners gather and collect data and information that can be used to explore alternatives and to compare business opportunities on a common basis. Financial and economic evaluations are conducted on best-available cost estimates to provide comparison for decision making. Table 1.1 provides some examples of financial indicators used in the decision-making process.

The internal rate of return (IRR) of a project is the computed annual return (expressed as a percentage) a project is likely to earn over the life of a project or investment. The IRR is also the discount rate that generates a net present value (NPV) of zero on a project or investment over the life of the project or investment. The IRR is one of the most common indicators used to assess the viability of a project. The threshold or hurdle rate of an owner or organization

TABLE 1.1

Common Financial Indicators Used to Compare and Select Projects

	Selection Criteria	
Indicator	High Number	Low Number
Internal rate of return (IRR): returns of a project expressed as a percentage	X	
Net present value (NPV): value of the project expressed in today's dollars	X	
Maximum capital exposure (MCE): Impact on the debt burden and organizational cash flows		X
Cash flow projections: provide opportunities for covering operating expenses and debt servicing	X	
Payback period: period (generally expressed in years) in which a project is expected to pay for itself (i.e., recover initial capital outlay)		X

is dependent on the risk tolerance of the organization. An organization that accepts a low IRR (e.g., 8%) may indicate a high tolerance for risk, whereas an organization that sets its hurdle rate at more than 12% may be considered a risk-averse organization.

Similar to IRR, the NPV of a project is a well-used indicator for assessing the viability of a project. Often, both the IRR and the NPV are used together in assessing the viability of projects. The higher the NPV of a project, the more attractive is the project. The NPV is an indicator of the value created by the project expressed in today's value. NPV is calculated by discounting the earning of a project over the life of the project to the present-day value. An NPV less than zero will result in rejection of a project. An NPV greater than zero may generally result in investment in the project. An NPV equal to zero may result in either acceptance or rejection of the project. In this case, other project criteria such as employment creation or other economic or strategic benefit must be considered in the decision-making process.

The MCE (maximum capital exposure) represents the maximum cumulative amounts of capital outflows to which the project exposes the organization. This generally represents negative cash flows to the organization and can be very intimidating for decision makers, particularly when the MCE is high and other project indicators are weak (low NPV and IRR). Approved projects with high MCE and other low indicators are generally projects with long-term returns and a long project life, in the range of 20–40 years.

Strong cash flow projections are an indicator of the robustness of the project or investment and an indicator of how quickly the project or investment may be able to recover its capital expenditures. In addition, strong cash flows are useful in tempering the impact of a high MCE and are an indicator of how quickly the project transitions from negative earnings to profitability. Projects with strong and sustainable cash flows are generally easy to select and approve.

Other nonfinancial criteria used in the selection process may include the economic impact on the region in which the project is to be situated, the number of obstacles to be removed to allow execution of the project, the strategic importance of the project to the core business of an organization, and even the political importance of the venture. These are generally useful for project selection when the NPV is low or zero or other financial indicators do not readily support the project.

In most organizations, approval authority may vary based on the strategic importance and capital outlays of the project. Projects that are required to address immediate personnel safety may require the approval of a line or facility leader only. On the other hand, if a project is intended to increase production or change the existing technology and large capital outlays are required, vice president (VP), chief executive officer (CEO) or even board of directors level approval may be necessary.

The identification-and-planning stage of the project cycle is also characterized by adequate, but smaller, capital injection aimed at carefully studying the unacceptable condition or the market opportunity. During this stage, an organization will seek to identify possible solution alternatives or best approaches for taking advantage of the opportunity. The organization will also seek to define the technology required to address the problem or to exploit the market opportunity. Alternatives are narrowed down to a few possible cost-effective and best economic options. At this stage, engineering cost estimates for a project are directional, with a ±30–50% variability in estimates.

Once a project obtains senior leadership approval to proceed beyond this stage, however, sufficient resources are applied to ensure that the project undergoes adequate rigor and scrutiny to generate confidence that the project will be successful. The project may be distributed into manageable segments to be worked on by smaller focus groups with resident expertise in the specific assignments before all the pieces are again pulled together to produce the final project. The urgency of this stage will often depend on the nature of the project. If the project is designed to eliminate an ongoing problem that has significant impact on production or returns, there is generally strong support to resolve the issue at the earliest opportunity.

If a project is designed to take advantage of growing lucrative market opportunities, the urgency may be much higher than that applied in the preceding discussion. Early market entry allows for desirable benefits of market share capture, such as price determination and cash cow opportunities. On the other hand, if the market is emerging or the technology is not yet proven, the project may proceed at a less hectic pace. If the potential opportunity relies on unproven technology that has not been commercially evaluated and capital cost is high, industry peers may seek investment partners to reduce risk, in which case a majority owner may be charged with the responsibility for executing the project. In addition, organizations may be hesitant to be market leaders since significant value can be derived from learning from the experiences and mistakes of early adopters.

It is not surprising, therefore, to see projects with long planning stages as organizations weigh and assess the performance of early adopters. Here, the focus is on learning from the mistakes of early adopters and capitalizing on their successes before proceeding. On the flip side, demonstrated successes by early adopters help to fuel the project pace, and information from industry peers and competitors is often well guarded.

Scoping and Design

The second important stage in the project management cycle focuses on putting together the design basis memorandum (DBM) and the engineering studies. The DBM is a detailed document that identifies the purpose, scope, stakeholders, and regional or economic impact of the project. During this stage, it is necessary to clearly identify the objective of the project, define the scope, determine its location, and define the principle of operation

FIGURE 1.2
Capital/labor intensiveness continuum.

from process and technological viewpoints. Here also, we expect to learn how the project will address the existing deficiency or capture the market opportunity.

The DBM will also clearly define the technology to be applied and will consider the location of the project, the scale of operation, the environmental impact, support services availability, and accessibility. Essentially, the DBM will define the basis on which the engineering design is developed. Types of equipment, mode of application, redundancy, operating pressures and temperatures, types of materials, and process resident time will all be developed during this stage.

Economic modeling and capital/labor intensiveness consideration will eventually define where the project sits along the continuum, with capital intensiveness at one end of the continuum and labor intensiveness at the other end (Figure 1.2). It is not uncommon to see economic considerations influence the choice of technology to be used in the project. The DBM has significant implications for influencing the size of the budget for implementing the project. During this stage, front-end engineering design (FEED) is completed. Here, it is extremely important to identify the *nice to have* from the *need to have* to prevent cost escalation and for successful project implementations and operations.

Careful selection of the project absolutes (need to have) must be undertaken since inadvertent elimination of a critical piece of equipment or component of the facility may have a significant impact on the operability and outcome of the project. This exercise requires the involvement of all stakeholders. All stakeholders must be reminded to carefully consider the business and operation critical needs of the project in defining needs versus wants. It is important to note that when this identification process is completed, the technology can change from a capital-intensive technology with low operating cost to labor-intensive technology with higher operating cost. Striking that happy medium between capital intensity and labor intensity along the capital/labor intensity continuum (Figure 1.2) is an important determinant in the overall operability and profitability of a venture. The capital or labor intensity is determined by the planned operating philosophy of the project.

A good example of how the capital/labor intensity decision is made is reflected in the choice between a fully automated capital-intensive sewer and drainage system and a semiautomated, labor-intensive system that must be manually emptied periodically. In making the right decision, care must be taken to give consideration to the impact on additional operating costs associated with third-party treatment, equipment maintenance, and operating philosophy. Similarly, in the absence of an effective sewer system, equipment preparation for maintenance can become difficult and costly as operating expenditure is increased in securing alternative methods for draining pumps, piping, and vessels followed by subsequent cleanup as the decision to shift from capital-intensive to a labor-intensive system is made. Cost can be escalated further based on the frequency of repairs, particularly if the equipment is not the best fit for the service required.

Execution (Implementation and Control)

The execution stage of a project is the transformation of a theoretical and planned concept into physical and material structures. During this stage, we see the actual site preparation, buildings, equipment, and machinery constructed (*systems*) and begin functioning to deliver the project objectives. The organizational framework and structure (*people*) are developed, personnel are hired and trained to operate the project, and adequate systems for managing and operating the project are developed. Also, *processes* such as procurement (supply chain management), maintenance management, accounting, and marketing are developed to meet the operational needs of the project.

During this stage of the project cycle, the project has the greatest potential for falling off the rails since both schedule delays and cost overruns generally occur in this stage. Adequate controls are required to avoid cost overruns or schedule delays. Critical to avoiding delays and cost overruns are the tools provided to, or developed by, the project leaders for early recognition and avoidance of these situations. The project leader, who now acts as the principal agent between the project and senior management of the organization, becomes the eyes and ears of the organization so that the right decisions are made at the right time with regard to the project.

Both the project cost and the schedule can be influenced by variables within and outside the control of the organization. Variables that fall within the organization's control include the approach to construction, labor productivity, site policies, safety culture, working conditions and wage rates, personnel turnover, level of training, and working hours, all of which can have a significant impact on the project schedule and budgets. Cost overruns can be further influenced by the organization's procurement policies, underestimation during the budgeting process, the absence of control and appropriate measures to ensure adequate control, and accounting procedures and practices.

Some variables external to an organization's control include environmental and seasonal conditions such as frigid winter temperatures, intense summer heat, and natural disasters such as tornados, earthquakes, and hurricanes. Nevertheless, astute planners will factor these considerations into the planning processes based on historical data and information. Sometimes, governmental intervention, regulations, and approvals can hinder progress and may be considered external to control. For smaller organizations for which the volume of business generated is insufficient to influence a supplier's behavior or when long delivery times exist in the supply chain management process, these factors may also be considered external to the control of the organization.

Closeout (Evaluation and Feedback)

Once construction is completed and the project enters the operations phase, performance evaluation and feedback are required to assess the long-term potential of the project. A comparison between actual performances versus planned performance targets is conducted. Such variables may include production levels, unit operating costs, and revenues. These comparisons are performed to determine whether the owner received what was paid for and to assess fitness for service of specific pieces of equipment, processes, systems, and organizational structure as a whole. Essentially, this is a performance evaluation or test of the plant or project.

This step is critical to the long-term success of the project. It is important to note that on schedule and on budget are not the only requirements for a successful project. Despite successful delivery of a project, in the absence of a capable and strong operating staff with the right processes and control systems in place, the task of achieving project targets becomes infinitely more difficult. Failure is imminent in the absence of a competent operating organization.

Reliability testing and performance evaluations are conducted to determine the integrity and capability of the project. Independent audits are conducted to determine gaps that have not been fully addressed. In most instances, this type of audit can take the form of a facilities readiness review (FRR) at various stages in the implementation process. An FRR focuses on the operational integrity and supporting requirements for sustaining operations of the project. An audit of this type will identify areas of weakness for the physical asset and organizational infrastructure. This audit will make recommendations for closing gaps identified, assign responsibilities for these gap closures, and suggest a timeline for closure. Often, an appropriate tracking mechanism is also provided to ensure responsibilities and accountabilities are met.

Adequate processes must be in place to ensure that design deficiencies, equipment failure, nonperformance history, and procurement processes are properly documented, captured, and addressed. Here, also, it may be appropriate to realign the organizational structure to ensure best personnel fit in assigned roles. In the simplistic form, this phase of the project cycle allows for a realignment of resources to address those areas of the project that prohibit the project achieving its designed targets.

2

Practical Workforce Management Tips

Introduction

During my experience in project management, I have made several observations that helped me to make the project execution stage of a project easier and more complete. These observations can be categorized into two main groups:

Group 1: Concerns observations relating to managing the workforce and people executing the project

Group 2: Related to observations associated with managing the physical assets and processes necessary for sustaining the project

This chapter reviews and discusses observations associated with managing the workforce. I also attempt to define the benefits of these observations and their contributions during project execution. The list is by no means all encompassing and can be developed quite easily based on the industry of application. The observations discussed here are applicable to industries such as oil and gas, petroleum, chemical, power generation, construction, mining, forestry, and most commercial steady-state facilities. In short, these observations apply across most industries and can be modified to suit prevailing circumstances. These observations are as follows:

1. Have the right person for the right job: Leadership skills are an absolute.
2. Make safety a priority.
3. Hire a mature workforce.
4. Place emphasis on developing strong teams.
5. Ensure clear expectations, responsibilities, and authorities are communicated.
6. Have a retention plan (avoid being a training ground).
7. Treat everyone fairly and with respect.
8. Encourage and support leadership visibility at the front line.

9. Embrace and promote diversity in the workforce.
10. Recognize and reward exceptional contributions.
11. Celebrate milestone achievements and successes.
12. Avoid conflict among operations, construction, and commissioning organizations.
13. Have an RACI (responsible, accountable, consulted, informed) chart.
14. Communicate, communicate, communicate.
15. Rotate personnel out as required.
16. Ensure adequate facilities are available.

These project management observations are now discussed in greater detail to provide a better understanding of how and why these observations are useful. Although applicable across most industries, focus is on larger industrial-type projects typical of the oil and gas and energy industries.

Have the Right Person for the Right Job: Leadership Skills Are an Absolute

Crucial to the success of any project are the skills and experience capabilities of the people who support the project. Here, the focus is on the skills and experience of key and functional leaders that are necessary to support the execution stage of the project cycle. A relatively flat organizational structure is desirable so that communication and decision making can be achieved swiftly without bureaucratic influences. However, regardless of organizational structure, the skills and capabilities of leadership personnel are critical for success in project execution. For all leaders, prior project management experience is required in any leadership role associated with implementing a large industrial project. In addition, flexibility to work long, demanding hours is an absolute requirement for assuming such roles.

Project and leadership personnel cannot be hired and be expected to learn on the fly during project execution. However, a high propensity for continuous learning is required for all leadership personnel in project execution roles. Experience in project execution is a unique characteristic, which starts with the acceptance that regardless of the rigidity of the plan, there will be the need to vary from it on occasion. Therefore, a strong ability to lead and manage change is required. Project leaders must also possess an excellent ability to multitask with decisive and strong decision-making skills. Excellent communication and leadership skills are also essential for assuming project leadership roles.

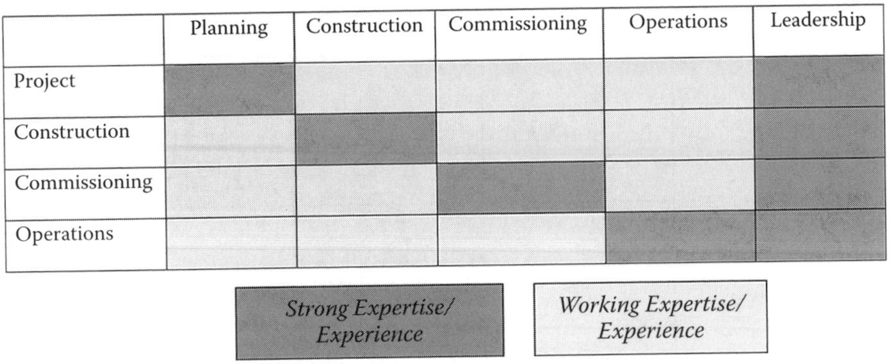

	Planning	Construction	Commissioning	Operations	Leadership
Project					
Construction					
Commissioning					
Operations					

Strong Expertise/Experience *Working Expertise/Experience*

FIGURE 2.1
Expertise and experience of project and organizational leadership personnel. (© Suncor Energy Inc. With permission.)

A good project leader must have gained prior experience across the three main areas in the execution of an industrial project. This includes experience in construction, commissioning, and operations activities. There must also be overlapping skills and knowledge by the construction, commissioning, and operations leaders. This overlapping expertise is best demonstrated in Figure 2.1.

While the project leader is not required to be an expert in all three areas, the leader must have a good appreciation of the challenges of leading associated with each discipline. Figure 2.1 shows the areas of strong and working expertise or experience of project leadership personnel. It is important to note that in most instances the required combinations of skills may not be readily available, and some learning on the fly will be necessary across all leadership disciplines. In addition to skills and working experience shown in Figure 2.1, all leaders must possess strong communication, technical, and leadership skills to function effectively in assigned roles.

The leadership environment has changed dramatically over time and continues to evolve daily as the workforce transitions from generation to generation. Today's leadership environment is characterized by high expectations from workers and high leadership efforts necessary to meet the expectations of the workforce. Figure 2.2 demonstrates the evolving leadership environments. In addition, the leadership skills required to lead and inspire the needs of a growing generation Y worker population (workers born after 1970) are very different from the leadership skills required for generation X workers. As a consequence, leaders who are resistant to change and fail to evolve with the changing workplace requirement are destined to fail during project execution.

Leaders must master modern leadership skills and inspire generation Y workers. At the same time, they must be able to cater to the needs of generation X workers in the same workplace. Meeting the different needs of both

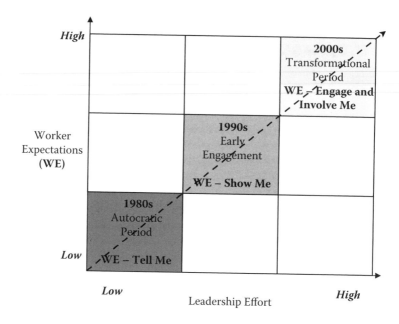

FIGURE 2.2
Recent evolution of the leadership environment.

work groups must be seamless and efficient. This is particularly important in the development needs of both work groups and in the way we train these workers.

Creating and sharing the vision for the project during the execution stage is critical for success. Workers want their efforts to represent something useful and productive. They all want to drive by a facility or structure and, with pride, say to their children, friends, and family: "I helped to build that plant." They do not want to be associated with failure. As a consequence, to retain workers and to keep them motivated, leaders must create a shared vision for the project. If the vision is shared, workers will remain and work on the project through completion. In the absence of a shared vision, the project becomes primarily a source of income until the next available job comes along.

Creating a shared vision requires leaders to demonstrate commitment to the project in a visible manner, and these leaders must have the ability to articulate clearly what the project is designed to do and how to arrive at that end. Communicating from the heart in an honest manner is also important. Followers will raise many issues during the execution stage that may contest the vision. Leaders must address these concerns in a practical and timely manner and take the time to address the concerns of followers. Failure to do so will result in reduced confidence among the workforce and a vision that may not be shared by all.

Keeping workers motivated is a continuous challenge for the project leadership team. Within their respective organizations, the construction, commissioning, and operations organizations may find unique ways to keep their workforce motivated. Fair treatment, involvement and engagement, challenging work, and career development opportunities are all methods that are available to the project leadership team for maintaining a motivated workforce. Other, less-recommended methods for motivating workers may be wages and salaries. In my experience, however, the effect of wages and salaries as a motivator is short lived, and workers become focused on the next wage increase to be motivated for another short period.

Demonstrated honesty and integrity by project leaders is an absolute requirement in the project environment. Transparent consistent behaviors, ability to do the right thing, and ability to recognize and acknowledge errors and mistakes all contribute to the honesty and integrity of project leaders. Followers are always looking toward leaders to understand how decisions are made and to emulate these behaviors. In the absence of honest leaders with integrity, it can become difficult to operate within the project execution environment, and workers with true values and integrity will leave.

Trustworthiness among leaders during the project execution stage is essential for success. Trust is an earned entity. Trust is earned by leaders demonstrating behaviors consistent with ethical and right decision making. The way leaders operate in the field will determine how much trust they receive from the workforce; unethical, irrational behaviors, dishonesty, unfair treatment to workers, and the way leaders interact with and treat workers will determine how much trust they receive from workers.

Project leaders, like leaders in other areas of a business, must be able to demonstrate empathy and care for workers. To do so during the project execution stage can be perceived as more difficult because of schedules and cost constraints. However, project leaders must address worker treatment in the same manner they would want to be treated. The way leaders treat workers will determine how trusted they become. For example, when leaders allow workers to respond to family crises as a priority over project schedules, trust is enhanced. Project leaders must then find innovative and creative means for maintaining a schedule when critical workers are released to address problems they may deem important to them and their families.

Team-building skills are important to project leaders. Such skills are required to develop and maintain strong teams. When strong teams are developed, productivity is increased. Workers are motivated, and schedules and cost targets can be met. Teamwork allows targets to be met even in the absence of a team member who may be called off for short periods. Project leaders must be skilled in team-building techniques and must seek to meet the team-building needs of an often diverse workforce.

Equally important in aligning people with roles is the need to have frontline leaders with strong leadership skills and capabilities. Figure 2.3 shows the key skills required by frontline leaders for each organization during project execution. Frontline leaders must be equipped with leadership skills and behaviors that are consistent with the owner's expectations and needs.

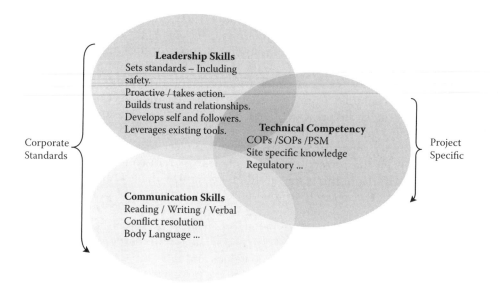

FIGURE 2.3
Skills requirements of frontline leaders. (© Suncor Energy Inc. With permission.)

As such, they must be appropriately trained so that they are aware of the leadership expectations.

Similar training is required to ensure frontline leaders are also technically competent and can lead the workforce in a competent and confident manner. During project execution, followers want the reassurance that leaders are doing the right things and that they are competent in what they do. Training of frontline leaders helps to cement corporate standards for leadership and culture and to ensure that they are adequately knowledgeable in the technical and regulatory aspects of the project. Technical training should be generally site or project specific with full knowledge of corporate standards such as codes of practice (COPs). Technical training must also include training on standard operating procedures (SOPs), process safety management (PSM) standards, and regional, provincial, or country regulations.

Beyond having the right person in the right role, there is a requirement for an organizational structure that supports the project execution stage. The project organization is made up of construction, commissioning, and operations organizations as shown in Figure 2.4. The project organizational structure must evolve with the project needs and must adequately meet the needs for sustaining the three organizations involved in the project execution in a phased sequence. Accountabilities and responsibilities for each role must be clearly defined. At some point during the execution stage of the project, the construction, commissioning, and operations organizations will coexist and work together. An organizational structure that supports this evolution

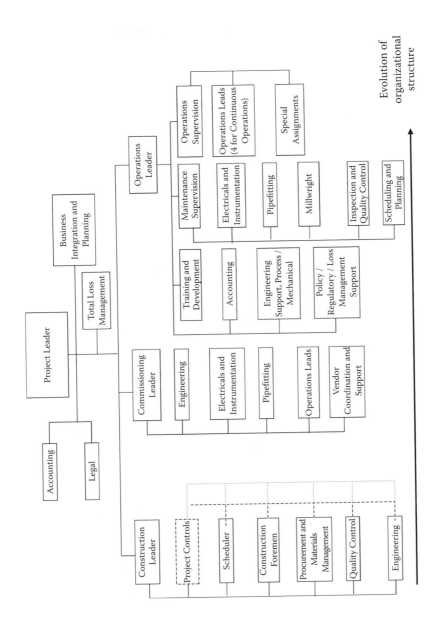

FIGURE 2.4
Organizational structure to evolve during the execution stage.

is required. Sequentially, a project leader should expect the organization to evolve in the order of construction, commissioning, and operations. Project leaders during project execution include the project leader as well as the commissioning, construction, and operations leaders. Other project leaders will be designated by the leaders of these organizations.

Make Safety a Priority

Leadership is critical in establishing the desired safety standard during project execution. The project leader is responsible for this. The overall safety standards and achievements of any project are important to all stakeholders. Poor safety standards and performance may result in people being injured or killed on the job. Damage to equipment, machinery, and the environment is also possible due to poor safety standards and a weak safety culture. Making safety a priority by the owner is important for establishing the eventual safety culture at the facility.

Adherence to safe work practices, the use of personal protective equipment (PPE), adherence to established COPs, use of SOPs, and the adoption of behavior-based safety training and standards at the project site all contribute to worker safety at the project site during execution. Each project organization may have one or more dedicated safety experts for ensuring safety in the workplace. There must also be close alignment among the safety standards of all three groups to avoid confusing workers.

Demonstrated safe behaviors by representatives of the owner set the standards for contracting workforce and personnel while they work at the project site. If a contractor or temporary employee sees an owner's employee not wearing safety glasses at the work site, the contractor or temporary employee may be inclined to emulate this behavior. Before project execution work begins at any project site, representatives of both the owner and the contracting organization must collaborate on the safety standards and procedures to be adhered to during project execution.

Setting safety standards, monitoring and addressing unsafe work behaviors, and continuous safety training are essential for safety outcomes and ensuring that the safety standards during project execution are met. Project leaders must clearly communicate to all personnel that safety is the responsibility of all workers at the site, and unsafe work behaviors will not be accepted or tolerated. Leaders must also follow through on consequences for deliberately ignoring safe work practices and continually strive to ensure the safety of all workers, assets, and the environment. A firm belief by all that zero harm to people, assets, and the environment is not only possible but also achievable must be shared by all organizations and stakeholders sharing the work site. Such a belief results from demonstrated safe work behaviors of all site leaders and zero tolerance to unsafe work practices and behaviors.

Project leaders must establish and steward both leading and lagging safety indicators. Leading indicators are proactive measures adopted by project leaders to help prevent incidents before they occur on the work site. Leading indicators do not prevent incidents; however, they provide early indicators for potential incidents that can be acted on by project leaders to mitigate against the occurrence of incidents. For example, monitoring of near misses provides project leaders opportunities to proactively remind the workforce via safety stand-downs of the potential impacts of major near misses and for their support in working safely. Similarly, the numbers of planned and unplanned leadership visits to the front line also provide leaders opportunities to engage workers informally and formally to better understand the safety concerns of workers. In addition, leadership visits to the front line allow leaders to observe workers at work and provide opportunities to address observed unsafe behaviors and reward safe work behaviors. Conducting safety audits is another powerful leading indicator that can enhance the safety standards of the project environment.

Halama et al. (2004) (of the Construction Owners Association of Alberta, COOA) listed 300 leading indicators in workplace safety. Halama et al. reduced these 300 leading indicators to the following top 10 leading indicators (p. 26):

1. Behavioral Based Observation Process Is In Place and Working.
2. Focus Observation Process and Focused Inspection Process Is In Place and Working.
3. Near Miss/Near Hit Reporting Process Is In Place and Working.
4. Employee Perception Surveys Are Conducted To Determine State of EH&S Health.
5. Pre-Hire Screening of Employees (D&A) Is Conducted.
6. Contractor Selection (EH&S) Process Is In Place Prior to Start of Project.
7. Active Management Safety Participation—Tours/Walkabout/Written Communications.
8. Supervisor Safety Activity Evaluated.
9. Hazard ID/Analysis Process Is In Place Prior To Start of Project.
10. FLRAs [field-level risk assessments] Are Conducted Prior To Start of New Work/At The Beginning Of Shift.

Lagging safety indicators are those that provide project leaders a look into the rearview mirror on safety performance at the project site. Lagging indicators may include injury frequencies such as total recordable injury frequencies (TRIFs), employee recordable injury frequency (ERIF), and contractor recordable injury frequency (CRIF). Other lagging indicators may include the number of times first aid is provided, number of loss time incidents, loss time injury frequencies, number of workers on restricted work, and the amount of loss time incurred by employees injured and away from work.

The key to success as far as safety is concerned is that project leaders regard expenditures in safety as a profitable investment with high rates of return. Essentially, such expenses are regarded as investments in the safety culture of the organization. When workers understand that leaders genuinely care for their safety, morale and performance can be sustained at continuously high levels. Good safety practice requires leaders to use both leading and lagging indicators in the project site. At the base of it all is the need for *shared learning, continuous training,* and *competency among the workforce.*

Celebrating milestone safety achievements, tracking and communicating injuries, injury frequencies, and other leading and lagging safety indicators send the message to all workers that safety is important at the project site and that workers must work safely. An effective risk management and risk mitigation process for each task must be available to assist workers in working safely. The direct and indirect consequences of unsafe work on worker morale and performance *cannot be underestimated,* and all workers must work diligently to ensure the safety of each other. Ultimately, *safety is everyone's responsibility.*

Safety and the Contract Workforce

Historically, contractor safety performance lags employer safety performance when frequencies are measured. In my experience, injury frequencies for contractors can range from two to four times the injury frequency of employees. Safety of the work site must address the safety needs of both employees and contract work groups alike. Often, owner organizations leave the safety of the contracted workforce up to the leadership of the contracting organizations. When an incident occurs on a work site, the impact does not discriminate between employees and contractors. Both employees and contract workers gravitate to low performance as these work groups deal with the impact of the incident. Morale and work performance are affected in both groups. Recovery time varies between groups and can significantly affect productivity.

The key, therefore, is to ensure that leaders maintain an effective contractor management process with one element focused exclusively on the safety of contractors. Many organizations recognize that managing the safety of contractors is a two-stage process. Stage 1 of contractor safety management focuses primarily on screening contractors carefully to ensure that only the safest contractors are allowed to work at an owner's site. Of course, to attract the services of a safer contractor, the owner may likely incur a higher cost since creating and maintaining a safe work environment require sound investment in safety in the form of training, vigilance, reporting, and data management of leading and lagging indicators. Nevertheless, leaders must recognize the trade-offs of incurring this cost up front as an investment in the project that yields benefits later—higher morale, fewer injuries, and direct associated cost and fewer incidents of modified work, which result in up to 40% reduced productivity for the same hourly rates as injured workers are reintroduced into the workforce under the umbrella of modified work versus a cheaper contractor who sacrifices safety for cost.

During the contract award process, therefore, leaders and stakeholders must maintain an effective contractor qualification process that assesses the historical safety performance of the contractor. There must be close alignment between the owner's safety philosophy and that of the contractor to maximize synergies and benefits from a safe workplace. Stage 1 therefore focuses on excluding or minimizing the safety concerns from the work site through careful screening and selection of contractors. This can be achieved by a series of questions asked of the contractor to determine the contractor's attitude toward working safely as well as the contractor's previous safety history.

When specialized services or work are necessary and contractors do not meet the safety standards established during the selection process, both owner and contractor must work together to develop an effective mitigation plan to ensure work is completed safely. Senior leaders must be aware of the mitigation plan and agreed consequences for variation from the mitigation plan, and performance standards must be followed through to maintain credibility and focus on commitment to safety.

Stage 2 of the contractor safety management process recognizes the need to manage safety performance when contractors are brought in to the work site. Stage 2 recognizes that regardless of the historical safety performance of contractors, there are leadership behaviors and focus that help to sustain the safety performance of workers and contractors alike.

Stage 2 requires leaders to inspire the hearts and minds of all workers to work safely. To achieve this goal, leaders of both the contracting and owner organizations must be visible at the front line to meet and talk with front-line workers. Leaders of both the owner and contracting organizations must be active listeners to genuinely *listen* to the concerns of workers. They must listen with genuine empathy to the concerns of workers where safety is concerned and act on them as necessary. Owner leadership visibility must not be limited to the owner's site only. If work for the owner is being done at the contractor's facilities, the owner's leaders must also be visible at the contractor's site to understand the behavior changes required between the contractor's work site and the owner's work sites. The fewer the differences in safety behaviors at each site, the easier it becomes to manage the safety performance of a contracting workforce.

Critical to success of stage 2 contractor safety performance management is the need to create a *no blame* culture. Here, the focus is on encouraging contractors to report near misses and leading indicators so that proactive action can be taken before an incident occurs. Owners must recognize that contractors may fail to report severe incidents and near misses if punitive actions are likely. Leadership must develop short- and long-term trusting relationships with contractors to ensure that all incidents and risks are reported in a manner consistent with the owner's reporting standards. Open, honest communication and the credibility derived from demonstrated behavior are essential for creating and sustaining a no blame culture.

Managing the safety performance of contractors require leaders to treat all workers the same as far as safety is concerned. The use of consistent safety procedures for both contractors and workers, standardized PPE, and similar training and access to information help us to blur the difference in treatment between contractors and employees.

Making contractor safety performance a key performance indicator for both owner and contractor is a powerful tool for enhanced contractor safety performance. What gets measured gets done. Strategic focuses on realistic and achievable safety standards help drive both contractor and owner behaviors toward a safer work environment.

Finally, before concluding this section, I must mention the use of assigning prime contractor responsibilities as a strategy to transfer safety responsibilities from owner to contractor. From my perspective, this strategy serves only as a means for playing with numbers. When a worker is injured or killed within a work site where prime contractor responsibilities are demarcated by a plastic fence, the impact is similar on the entire site workforce. Similar vigilance and safety standards are required for both owner and contractor to maintain a safe work site regardless of prime contractor status on a common work site.

Hire a Mature Workforce

While it may not always be possible to have a mature workforce or desirable for the long term to have a mature workforce, during project execution a mature workforce is an extremely valuable and desirable requirement. The maturity of a workforce is reflected in the skills and experience of the workforce, and there must be a greater number of experienced and skilled workers than inexperienced and unskilled or new workers across the construction, commissioning, and operations organizations.

The project execution period is not the time for personnel in leadership roles to be learning to execute projects. However, I must point out that it is by far the best time for inexperienced workers to learn about the three major elements of project execution: construction, commissioning, and operations. For inexperienced personnel, the ability to crawl into vessels to examine internals and visualize how a system works is an invaluable learning experience. For each year of work performed during project execution by new and inexperienced workers, this will represent on average 4–6 years of equivalent work experience on a similar steady-state facility. Operating and troubleshooting skills are also enhanced during this period.

Nevertheless, the workload demands during this period are so high that it is extremely difficult for personnel to be focused on learning as opposed to delivering on people, process, or system readiness deliverables. In the absence of a largely mature workforce, experienced workers are stretched to the limit to deliver work and to provide hands-on practical training to junior, less-experienced employees. A dedicated training officer can help to develop inexperienced workers during this period.

In heated market economies, intense competition may lead to a paucity of skilled labor. During such periods, mature and skilled workers may be difficult to find, leading to intense competition and inflated wages and salaries for workers. It may be necessary to incur this additional cost to ensure the project gets the best opportunity for achieving the goals of schedule and budgets. In the end, this additional cost will be outweighed by the benefits derived from delivering the project safely and on schedule.

Place Emphasis on Developing Strong Teams

Strong teams are required in all organizations during the project execution stage. The project leadership, construction, commissioning, and operations teams must all be highly developed to meet the requirements of the fast-paced execution stage of the project. Team building must consider organizational, cultural, and technical fit to generate a well-balanced, highly motivated team to meet the deliverables of the team.

To meet the challenges of a strong team, leaders must be provided both flexibility and resources to promote team building within organizations. In addition, leaders must be trained in techniques for building strong teams. We have all heard the expression that "a team is only as strong as the weakest link." This statement is exceptionally true in the execution stage since the weaknesses and failures of one team member can result in disastrous consequences for both project cost and schedule.

Clark (2008) advised that, regardless of industry or company size, strong performance is derived when workers see themselves in the same boat, rowing in the same direction toward the same destination. Clark also advocated team commitment and placement of personnel in roles in which their strengths can be exploited. Building strong teams requires team leaders to know each member of the team, understand their strengths and weaknesses, and leverage their strengths while developing them to eliminate weaknesses. As a consequence, building the competency of a team is hard work and requires considerable efforts from team leaders. The goal of every team leader is to develop a team that can function effectively in the absence of any member until the void left by a team member is refilled.

To create strong teams, team leaders must provide sufficient information to team members to support their climb on board. The leader must create a shared vision. Leaders must also create a compelling sense of direction to all members of the team and use various skills to motivate them in this direction. Essentially, the leader must create and share a road map with workers on how the *team* shall get to its destination. Finally, leaders must clearly define the destination and provide continuous updates on progress toward this destination.

In creating strong teams, the skills of a transformational leader cannot be underestimated. The actions of a team leader must create trust and generate confidence among the team under his or her leadership. As far as teams are concerned, team members are prepared to forgive failures or disappointments once or twice. Occurrence more than two times tends toward a breach of trust among members and leaders. Team members and leaders must therefore work hard to earn and retain the trust of each other in a team environment. Fair treatment is also an essential ingredient in developing strong teams. Individual consideration is also essential in developing each worker for the role he or she is required to perform. Strong communication and supporting skills are also required to ensure that each member of the team is aware of his or her role and how this role has an impact on the overall team performance. Strong supporting skills are required to develop each member of the team to optimal performance for the role each may hold in the team.

Beyond the development of team members within the work environment, leaders must seek to promote team building beyond the confines of the work environment. Family picnics, as an example, help in promoting social bonding of families and a closer integration of team members in the work environment. Such events also promote leaders as caring individuals and increase the cohesion among team members and eventual performance.

Ensure Clear Expectations, Responsibilities and Authorities Are Communicated

Expectations relating to the organizational values must be established very early for all employees who work at a project site during execution. Such expectations may include safety standards and safe work practices, respect for each other, zero tolerance to drug and alcohol usage, honesty and integrity and bringing mistakes to the attention of leadership early such that they can be corrected before they become larger problems at a later date. Expectations must be communicated also in the behaviors of peers and leaders, both of whom must demonstrate in their behaviors the behaviors expected of new employees. For example if safety is a priority, leaders and peers must not be seen without required personal protective equipment, nor should they be seen participating in risky behaviors or performing work without a pre-job hazard analysis and risk mitigation plan.

During project execution, efficiency of efforts and removal of ambiguity in the roles of all workers is essential for the safe and scheduled completion of work. Overlapping responsibilities increase the possibility of work being left incomplete as each worker may likely leave the incomplete task to the other when many competing tasks are to be completed. Similarly, gaps in responsibilities can also result in work or tasks being incomplete since the worker may not be aware that this is a part of his or her responsibilities.

Leaders must seek to ensure that all workers are made aware of their responsibilities and duties to ensure work gets completed on schedule. This means, therefore, that functional leaders must be engaged in defining the roles and responsibilities of workers and until a worker is mature and competent in the role, frequent contact between leader and follower is required. Moreover, in the fast paced project execution environment, leaders must strive for daily interface with followers to ensure obstacles to getting work completed are addressed.

Of equal importance to all employees are the levels of authority across the project. Authority provides opportunities for decision making and getting work done when obstacles are met and alternative approaches to the planned approach are required. Level of authority and delegated authority allow senior leaders to focus on strategic planning work while minor problems are addressed by lower level leaders. Full communication of levels of authority provides followers clear understanding of who can help resolve problems they may encounter during the course of their work. In addition, during emergency situations, knowing who to turn to for assistance can significantly reduce the severity of incidents and speed at which they can be addressed.

Have a Retention Plan (Avoid Being a Training Ground)

As discussed, heated market economies can lead to intense competition for skilled and experienced workers. Organizations will continue to compete for skilled and experienced workers. Workers also maintain an informal network and will often provide feedback to leaders and employees regarding the skills and capabilities of fellow workers and the benefits of working for one employer versus another, respectively.

This informal network is extremely important and can slowly bleed employers of skilled and experienced workers in the absence of an effective retention plan. To avoid this undesirable turnover, project leaders must develop creative means for retaining employees. To best retain workers, the project leadership team must understand the things that are important to the worker for developing long-term loyalty. In addition, it may be necessary to define the payback period to the project for cost incurred in hiring, training, and retaining the workforce to affect a good retention strategy.

Among the retention strategies possible are the following:

1. A retention bonus with a locked-in clause. Workers are provided a cash bonus up front in return for their commitment for a fixed period, such as 3–5 years. This method works well at least until the locked-in period has expired.

2. Ensure development opportunities are available for all employees. Project leaders must create a line of sight between the employee's current roles and where the employee is likely to be after given periods, say 3–5 years. If the employee is satisfied with this projection, the likelihood of retention is high. It must be noted, however, that project leaders must enable the employee to achieve these goals or turnover may take place.

3. Create a satisfying workplace for the worker. Research has demonstrated conclusively that in the absence of job satisfaction, workers will leave. The project leadership team must carefully determine the variables that maximize worker satisfaction in the particular project and work toward creating this work environment. Research has shown that leadership styles and behaviors, access to training and development opportunities, teamwork, individual consideration, empathy, listening to the employee, creating a safe work environment, cultural awareness and sensitivity, and fair treatment are variables that improve satisfaction in the workplace. When coupled with a competitive wage system, this approach to retention is very effective and sets the foundation for a progressive organizational culture when the project moves beyond the project execution phase.

We have often heard the phrase "workers do not leave jobs, they leave leaders." To avoid turnover, therefore, the skills of project leaders must be developed to ensure optimization of worker satisfaction during project execution. Project leaders must be assessed for weaknesses and gaps in their leadership competence and appropriately trained to close the leadership skills gaps.

Treat Everyone Fairly and with Respect

An important requirement for successful project execution is the need to treat everyone fairly and with respect. Failure to do so creates an environment characterized by high turnover and lowered productivity. Fair and respectful treatment is important since project leaders seek to ensure synergy among various work groups and personnel and to create an environment in which all workers are motivated to higher levels of productivity. In today's global and diverse work environments, project leaders will encounter workers from varying ethnic, religious, cultural, and regional backgrounds. These workers will possess diverse knowledge, work experiences, work methods, and skills. Harnessing this complex mix of knowledge, experience, and skills in a beneficial manner has the potential to generate excellent performance in the workplace. To do so successfully, project leaders must ensure that all personnel are treated fairly and respectfully while leaders exercise high levels of cultural sensitivity.

Respect and fair treatment in the workplace are also very important since work groups today can be highly diverse in terms of ethnicity, gender, and sexual orientation. Harassment, unappreciated jokes and comments, bullying, and other similar behaviors can have huge adverse cost and productivity implications for the organization. Beyond the immediate impact of lower motivation and productivity, if disrespectful conditions prevail in the workplace, there can be a significant impact on the organization's goodwill and its ability to attract and retain qualified, skilled, and motivated personnel. Furthermore, employees can and may consider pursuing human rights violations charges, which can be particularly damaging to the organization.

Fair treatment is rooted in the rules system that governs areas such as compensation, benefits, rewards, and a working audit system. Equal pay for equal work should be the governing philosophy for leaders as workers are hired and put to work. Failure to treat people fairly can result in low morale, high turnover, and lower productivity. Ultimately, the project suffers in the areas of both cost overruns and delays. Disenchanted employees can create conditions that can lead to significant losses to the organization when the attempt to start up and operate the plant occurs. In my experience, organizations have demonstrated that disenchanted workers will perform substandard work and sabotage in the form of failure to remove debris and tools from process piping and vessels when they are welded shut and bolted. At start-up, costly shutdowns and expensive repairs from damage incurred by critical equipment resulting from these conditions can occur.

The need to treat all workers fairly and with respect can influence the success of any project during the execution stage. Often, it may be necessary to augment the leadership skills and behaviors of the project leadership team with training to ensure that they are equipped with the behaviors and skills necessary for treating workers fairly. All site leaders must be introduced to, and be required to support, the core values and leadership behaviors of the owner.

Encourage and Support Leadership Visibility at the Front Line

An area in which leaders frequently fail during project execution is in being visible at the front line. Leadership visibility at the front line sends a clear and strong message to all workers that leadership is interested in the work the frontline worker does, in the safety of the frontline worker, and in learning from the frontline worker the issues and challenges faced at the front line. The frequency of visits should be developed based on a balance among cost, impact, and effectiveness of the visit.

In situations of an immature and inexperienced workforce, more frequent visits by line management are required to support the development of the worker. Line management has opportunities to recognize and correct inappropriate behaviors, provide guidance and support to inexperienced workers, and offer immediate feedback to workers if appropriate. Senior leaders, through scheduled visits and engagement of frontline workers, have opportunities to listen to workers regarding challenges and issues workers face. By acting on the concerns expressed by workers, leaders can then dramatically enhance motivation and productivity.

The key to successful visits by senior leaders is to plan the visit in advance. In doing so, the senior leader must communicate with the frontline leader to understand the challenges faced at the front line. Senior leaders must be prepared to listen to workers and do so with empathy and caring. Senior leaders must practice what are called *listening moments* (periods when the goal is to listen attentively to all workers to better understand the many concerns of all workers as opposed to recording of data and information). More important, when concerns and issues are raised by frontline workers, senior leaders must take necessary measures to address the concerns of the frontline workers.

On sites where contractors may be involved in the construction of a facility, a joint visit by senior leaders of both the owner and contractor provides an extremely strong motivator to workers of both the contractor and owner during site or facility visits. Such visits send powerful messages to all workers that both organizations are interested in the progress of work and in the concerns of the workers. During leadership visits, however, leaders must be careful to ensure full compliance to site policies and safety standards since failure to comply with site standards sends the message of a double-standard system that will eventually lead to failure during execution.

Embrace and Promote Diversity in the Workforce

Project leaders should seek to promote diversity in the workforce. Diversity introduces creativity and innovative solutions to challenges experienced during project execution. On one occasion, during the start-up of a facility, a ruptured line resulted in the release of process fluids that could have resulted in contaminated waterways and resultant ecological damage.

While engineers scrambled to put together an action plan that depended on procurement of a holding tank and accessories to transfer the process fluid into the tank, the released process fluids continued to flow into the environment. Given the remote location of the project, a wait time for delivery of more than 4 hours was expected. An enterprising foreign-trained engineer not initially consulted in the emergency response plan quickly assessed the situation and provided a simple and effective solution to the looming problem.

With available site equipment and resources, the engineers proceeded to guide the construction of a collection pond for the process fluids. A backhoe at the project site was used to dig a containment pond that was lined with pond-lining materials available at the site. Once the makeshift catchment pond was constructed and lined, a trench was cut by the backhoe to divert the released process fluids into the temporary storage area, thereby averting a potential ecological disaster. The entire process was completed within 1 hour. Once the immediate danger was averted, a more permanent containment and repair plan was developed. This engineer was experienced in developing creative solutions for resource-constrained developing countries in his home country.

Cultural and gender diversity bring wonderful experiences and creativity to the workplace during project execution. Women leaders in particular promote the much-needed skills and leadership behaviors of caring for people and empathy for the workforce. In addition, they introduce greater creativity by seeking knowledge from probing questions in a nontraditional work environment.

Recognize and Reward Exceptional Contributions

Recognizing and rewarding exceptional contributors is the single most important motivator in the workplace. Timely leadership action in recognizing and rewarding exceptional contributors creates a positive image about the leaders and motivates workers to higher levels of productivity. Recognition of workers can take many forms and varies from the simplistic letter of recognition to commendations from corporate leadership. Similarly, rewards vary from small tokens to large cash or noncash rewards. The extent of the recognition or rewards may depend on the magnitude of the contribution. Recognition and rewards may begin with an individual worker and extend to the entire workforce.

A positive recognition and rewards system can generate and maintain high motivation, productivity, and morale among the workforce. However, a system that appears partial can have the reverse effect among workers. Care must be taken by the project leadership team to develop and steward a fair system for recognition and rewards and to ensure proper administration of it. Responding to safety concerns is generally an area in which positive rewards recognition can be extremely satisfying and motivating to the workforce. Trinkets, safety challenge trophies, and team-building social events are simple rewards that can be extremely powerful motivators to workers for demonstrating the right behaviors in the conduct of their duties. Workers also recognize that employers care for their safety when safety concerns are addressed in a timely manner when brought forward. This serves to motivate them further.

Celebrate Milestone Achievements and Successes

At the start of project execution, milestone activities are identified, usually aimed at communicating significant advancement toward the overall project completion (milestones are discussed in greater detail in Chapter 3). Celebrating milestone achievements is not only an effective means of acknowledging the efforts of workers but also a means for communicating to all workers the extent of advancement made toward completing the project.

To the worker, more specifically frontline workers, this is recognition of their victory and a celebration of their achievements and contributions to the project advancement and completion. This victory is even more resounding when the milestone activities are achieved ahead of schedule or within or lower than the projected budget. The scale and extent of the celebration depend on the significance of the achievement and the magnitude of the success as well as the resources available to the project team.

Avoid Conflict among Operations, Construction, and Commissioning Organizations

Site ownership during the various stages of a project is also important. It has been clearly demonstrated that site ownership during the stages of a project should belong to the respective majority system holder during the execution stage of the project. Generally, ownership transitions from construction to commissioning and finally to operations as the project transitions from construction to steady state to continuous operation.

Ownership of the facility or project ultimately resides with the operations organization (owners). With this in mind, during the construction and commissioning phases of execution, the operations organization will generally seek to ensure that it takes away the best possible functional facility. The operations organization will seek to obtain a facility that operates with high reliability, that is easy to operate, and that meets the design outputs of the facility. This mindset leads the operations organization to actively seek out and influence the final outcome of the project after the scope has been finalized. Indeed, the operations organizations seek to establish a facility with the best functional (operable) capabilities. If possible, the operations organization will seek to include *nice-to-have* capabilities that may significantly exceed the capacities and capabilities of the *need-to-have* requirements of the project.

On the other hand, the construction organization seeks to provide a facility consistent with designs, approved drawings, and scope. Similarly, the commissioning organization generally focuses on making the process, equipment, and machinery work at least for the duration of their involvement with the project so that ownership can be transferred to the operations organization. The construction and commissioning organizations are motivated to complete the job consistent with design in the shortest possible time to move on to the next job. The construction and commissioning mindsets do not necessarily include the functionality or long-term operability of the unit or facility.

These different mindsets between the operations, construction, and commissioning organizations can lead to conflicts that can impede the schedule and result in escalating costs. While the operations, construction, and commissioning organizations are generally aligned on the schedule and budget deliverables of the project, only the operations organization is concerned with the long-term operability of the facility or project since the product they receive will be with them for the life of the facility.

Resolving conflicts between schedule, budget, and operability or functionality requires creative decision making among all organizations. A risk-based approach to decision making regarding scope changes should be adopted in deciding whether changes to the design can be accommodated. In this way, when change requests are made, such changes must be supported if the designed system has been identified to have an inherent risk to operability or functionality that is removed by the proposed changes.

Another approach to resolving this conflict is to place some form of ownership on the shoulders of the construction and commissioning organizations beyond the turnover of the project to operations for 1–2 years. In so doing, functionality and operability concerns now become shared across all organizations involved in project execution. Such joint ownership encourages all organizations to cooperatively consider all functionality issues for a resolution beyond the design limitations imposed by blueprints. Functionality and operability become the overriding deliverables as all groups work to ensure functionality within budgets and on schedule.

Have a RACI Chart

What is a RACI chart, and why is it important? A RACI chart is a simple chart that defines who is responsible for assigned work, who is accountable for the assigned work, who should be consulted, and who should be informed about the status of the assigned work. The worst nightmare of any project leader is to hear the following simple questions or responses during the investigation of any incident or situation involving improper execution of work:

1. I did not know I was responsible for that task.
2. Does anyone know who is accountable for this work?
3. No one consulted me (us) before this task was undertaken.
4. I was not informed that this exercise was taking place.

How can project leaders manage the process to ensure these uncomfortable situations do not surface during the execution stage of the project? With the multitasking requirements of all personnel during this period, some work will inadvertently be forgotten and will not receive the attention required in a timely fashion to keep the project progressing forward.

A simple prompt and visible mechanism to remind personnel of their roles and responsibilities helps in avoiding these situations. A RACI chart serves to do exactly this and to assist project leaders in maintaining continuity across all groups by clearly identifying the roles and responsibilities of all personnel involved in moving the project forward. More important, the RACI chart can be applied for smaller activity management during the execution stage of the project. The acronym RACI denotes the following: responsible, accountable, consulted, and informed. Figure 2.5 represents a typical RACI chart identifying the roles and responsibilities of key stakeholders identified in the project execution stage of a commercial project.

A RACI chart is useful in eliminating confusion among personnel in terms of roles and responsibilities and generally serves as a powerful reminder to responsible and accountable participants on their deliverables. A RACI chart is an informal contract among individuals within a work group or across work groups. A RACI chart is extremely important in preventing duplication of work and avoiding work being inadvertently forgotten or incomplete. During the project execution stage of a project, many RACI charts may be employed for different work activities. This simple tool works well in supporting continuity in work and ensuring that work continues at a managed pace.

R—Responsible

Personnel assigned as "responsible" are responsible for the particular activity and will be required to ensure the activity is fully resolved and managed to completion during the project execution stage in the time allocated for doing so. Responsible individuals ensure that resources are obtained for completing assigned work.

A—Accountable

Personnel assigned as accountable are ultimately accountable for each assigned activity and will be required to ensure adequate attention, priority, and resources are allocated to the activity to allow the responsible person to properly execute the task. Generally, accountable individuals or personnel are the leaders of the three organizations on whose shoulders the final accountability for completion of any assigned work rests.

C—Consulted

All personnel assigned as consulted may have information useful for resolving or addressing those activities for which they must be consulted. The onus is on the responsible person in seeking the input of the groups to be consulted before the activity can be deemed complete.

I—Informed

Personnel designated as informed are included in the RACI chart for they are made aware of the status of activities and for decision making within their respective groups. Informed personnel have the opportunity to mitigate within their work groups against any unsafe conditions that may arise during the particular activity. They can begin planning for the next activity in the queue.

It is important to note that two different groups cannot be responsible and accountable for the same activity at any given time. To demonstrate how the RACI chart works, consider its application on the start-up of a typical steam-assisted gravity drainage (SAGD) facility. Figure 2.5 (RACI chart) provides a matrix of activities and personnel or work groups with RACI requirements for each activity.

	Project	Operations					Commissioning				Construction	
Named Representative	Project Leader	Leader	Engineering Support	External Services	Shift Team Leads	Room Operator	Leader	Engineering Support	Leads	Leader	Engineering Support	Leads
Activity				Jane	4-Leads	Operators			4-Leads			4-Leads
Green light to start	A	R	C	I	C	I	C	I	C	I		I
Morning Meetings – Readiness Assessment	A	R	C	I	C	C	I	I	I	I	C	I
Respective Organizations Informed	A	R	R	R	R	R	R	R	R	R	R	R
External Stakeholders Informed	A	C	I	R			C			C		
Progress Updates Issued Daily	A	R	C	C	C	C	R	I	I	I	I	I
Equipment/Design Deficiencies	A	C	C	I	C	C	I	C	C	C	C	C
Commissioning Related Upsets	I	R	C	I	C	C		C	C	I		
Process Related Upsets	I	R	C	I	C	C		C	C	I	C	
Commissioning Activity Completion	I	I	I	I	I	I		C	R	I	C	I
Specific Operations Activities												
Water Balance Recovery		C	C		R	C	I					
Fuel Purchases				I	R	C	I					
Communications – Running Instructions			C		R	C	I					
Scheduling and ensuring personnel availability			C		R	C	I					
Sample testing frequency and auditing			C		R	C	I					
Reporting / Correcting upset conditions	I		C	I	R	C	C	C	C	C	C	C
Specific Commissioning Activities												
Activity start	I	I	I	I	I	I		C	C	C	I	I
Activity Stop Success/Failures	I	I	I	I	I	I		C	R	I	I	I
Activity Delay	I	I	I	I	I	I		C	R	I	I	I

FIGURE 2.5

Typical RACI chart.

Communicate, Communicate, Communicate

Communication during the project implementation phase is an important requirement. Communication among and within groups is required to ensure safe and successful project execution. Communication generally takes place daily on a project site via a combination of methods, and in different forums, among manageable groups. Daily communication is required in response to the daily changes taking place on a project site during the execution stage.

It is extremely important that all groups share information daily about changing work activities, hazards, visitors' presence, introducing and orienting new employees, and progress updates. Communication starts at the beginning of the workday and continues throughout the day as work proceeds. Some of the more effective and common communication approaches on the work site include the following:

1. Leadership meetings
2. Toolbox and prejob meetings
3. Town hall-type meetings
4. Notices
5. E-mail and electronic communication
6. Combinations of all communication methods

Leadership Meetings

The leadership meeting at the start of each workday is probably by far the most effective means of communicating workplace concerns to all groups and stakeholders at the work sites. Figure 2.6 provides a simplified version of the interaction that is required between and among various stakeholder groups and work teams. Representation from various stakeholder groups is required to provide necessary high-level status updates and information relating to planned work and activities.

Leadership meetings must follow a structured format and agenda. Focus should be on reviewing achievements of the previous day and planned activities for the current day. When situations of near misses and safety incidents are experienced, these items must be elevated to the top of the agenda such that the philosophy of personnel safety first is a priority in every leader's work plans.

Toolbox and Prejob Meetings

Toolbox and prejob meetings are held among workers within specific work groups to provide information on assigned tasks for a particular job. Toolbox and prejob meetings are generally held at the start of a job task and will often

FIGURE 2.6
The communication and stakeholder interactions during project execution.

include the leads, foremen or -women, and those performing the job. At a tool-box or prejob meeting, discipline leaders will communicate planned activities as well as all high-risk and high-priority work that will be undertaken dur-ing the day. Specific details regarding each task are reviewed with workers performing the work. Risk mitigation plans, conflicting and concurrent work, reallocation of resources, rescue plans, and other detailed work practices are reviewed at this meeting between work group leaders and the field and skilled worker. It is also the opportunity to review recent incident circumstances and to discuss key messages coming from the project manager and team.

Two-way discussions at the toolbox or prejob meeting provide discipline leads with challenges and concerns regarding assigned tasks that will be fed back to the project leadership team on daily planned activities. Work progress, delays, schedule gains, and other decision-making details are fed back through discipline leads and supervision to the project leadership team. This two-way information flow is a continuous process and takes place via telephone, two-way radios, e-mail, or direct communication at the work site.

Town Hall-Type Meetings

Town hall-type meetings are designed for passing on information to all site personnel or large mixed work groups at the same time. The goals of town hall meetings are to ensure that all personnel receive the same message at the same time. Town hall meetings are often used to communicate successes, incidents, near misses, and other important information to the entire workforce at the work site. Generally, the project leader or a delegated site leader with the support of the project leadership team may be required to deliver the particular message. Town hall messages are generally precise and powerful and are designed to drive home a key message over a few minutes. An example of an effective town hall message is the reinforcement of work safety behaviors when a milestone safety achievement is surpassed, such as 1 million man-hours loss time free.

Notices

Notices are important forms of communication in the execution stage of a project. Notices are a highly effective means for communicating information to work groups, particularly to workers at the front line. Notices may be prominently posted where workers gather or congregate, such as lunchrooms and canteens. The impact of notices is heightened when they are discussed at toolbox meetings before they are posted. Greater details can be provided in notices for those who want to learn more about the particular issue.

Effective notices may include such information as hazards associated with systems, live systems, changes in work processes, and other information relevant to the health and safety of workers. Notices must be able to catch the attention of workers and should be concise enough to get the desired message out to frontline workers without taking too much time. Generally, a graphic with three to five key points should be included on the notice for greatest impact. Messages must be clearly stated and easy to understand. The graphic should also be accompanied by a few simple words that explain the graphic.

E-mails and Electronic Communication

Within this group of communication tools resides the use of e-mail, telephone, faxes, shared network drives, or varying combinations of these. Careful attention to the type, content, and audience of the message is required to avoid miscommunication and confusion. Progress updates must be precise and accurate, particularly since the project leaders are required to provide feedback to corporate leadership, and it is based on this feedback that corporate communications may be generated. Common knowledge information can also be stored on an open electronic folder that is *read access only* for those who may be interested in learning more about the project. Conversely, all sensitive and privileged or confidential information must be appropriately secured and stored.

Combinations of All Communication Methods

One area in which the best results are generated by a combination of all of these communication methods is in creating awareness regarding the introduction of new or process hazards into the work site. Informing workers of process hazard introduction may best be achieved using a combination of toolbox meetings, e-mails to leaders of work groups, signage, and bulletins on notice boards. A town hall meeting that gathers the entire workforce is also effective when communicating process hazards introduction. Full communication of this type of message using multiple communication methods is essential to minimize the frequency of near misses, incidents, and accidents. Figure 2.7 provides a sample notice and the details required for a new hazard introduction that must be communicated to the workforce.

All site leaders are responsible for ensuring that messages reach their intended audiences. Site leaders are also responsible for ensuring that messages are fully received by providing opportunities for those who need further information avenues to do so. In most instances, a single point of contact is required to maximize the quality of communication.

The importance of communication cannot be understated during the execution stage of a project. Clear, concise, effective, and complete communication at the project site is extremely important for the overall success of the project. A wide range of communication tools and methods must be adopted by project leaders to ensure all stakeholders are informed regarding the project status and activities: Communicate, communicate, communicate.

Rotate Personnel Out as Required

As project execution continues, depending on the workloads of personnel, the potential for burnout can be high. Careful project leaders will be continually vigilant for signs of burnout. When burnout is apparent, workers must be rotated out to continue to maintain high levels of productivity. Burnout

New Hazard Introduction # ___

Date: Day / Month / Year

SYSTEM OWNER: Commissioning

System #:	1	**System Name:**	**PERMANENT POWER**
Contact Person:	John Doe		
Telephone:	( )–( )–( )		
Issued By:	Commissioning		
Representative:	Tom Jones	**Risk Factor:**	High

Hazard(s) Introduction	**Reference P&IDs and Major Equipment, Systems Affected**
Power introduced from main substation and distributed to Motor Control Centers (MCC's) 1-6	See Drawing Number C1 211-E-100-106 and system equipment list.

SYSTEM IDENTIFIERS
Main incoming switchgear feeders from Substation Circuit breakers in MCCs 1-6

POTENTIAL HAZARDS	**PRECAUTIONS**
Electrical shocks / electrocution	-All 5KV feeder breakers are locked in open position -All 5KV motor breakers are locked in open position -Any work on this system requires a SAFE WORK PERMIT from system owner -All rotating equipment power supply is locked in the open position
Accidental equipment starts	

FIGURE 2.7
Sample hazard introduction notice. (© Suncor Energy Inc. With permission.)

may occur among leadership personnel when ownership tends to be high and there is a tendency to work long and continuous hours.

Poor design and difficulties encountered in moving a project into steady-state continuous operation can result in situations in which personnel can enter a *firefighting mode* (one process upset is followed by another and another and so on) and leaders are unable to think strategically to address root causes of process problems. If left unattended for a long period, this situation can result in high personnel turnover and a subsequent loss of valuable intellectual property by departing personnel. More important, the impact of departing trusted leadership personnel can have an impact on the morale of the entire workforce.

Often, the introduction of new personnel will result in new energy injection into the project and can revitalize the leadership team. Such changes may introduce a fresh perspective on the project. Proper transfer of responsibilities and knowledge is required during the rotation of personnel. A drawback of new personnel introduction, if not planned properly, is the creation of a steep learning curve to the incoming personnel when information transfer is incomplete or inadequate. Such a situation can spiral into declining productivity and greater losses. To avoid such conditions, it is important that outgoing and incoming personnel are allowed to properly transition through an effective job shadowing process in which new personnel work alongside outgoing personnel for a period of 2–3 weeks before transition is completed. The outgoing personnel should not perceive such actions as punitive.

Ensure Adequate Facilities Are Available

During the project execution stage, the amount of new workers increases at the project site on a daily basis. Care must be taken by project leaders to ensure that adequate support facilities are available to meet the needs of the growing workforce. *Facilities* refer to accommodations, office space, meeting rooms, kitchen facilities, washrooms and toilets, copiers, telephones, presentation equipment, and all other equipment required to make the day at work a pleasant and fruitful one.

As project execution continues, temporary accommodation must be made available to accommodate a transient workforce that is necessary to support execution activities but will not remain within the operations organization on a long-term basis. Between the start of project execution and the point at which the operations organization (the final owner) takes ownership of the project, the manpower loading on the project site undergoes the transformation shown in Figure 3.2. As such, the project must cater to a larger workforce than the one that will eventually inherit the project and its facilities.

Shortfalls in facilities can be accommodated through rental services or construction of temporary facilities. Personal amenities are important during this period. Proper kitchen and eating accommodations and adequate, well-maintained washroom and toilet facilities are required to maintain a comfortable, productive, and disease-free work environment. In the absence of adequate facilities, personnel frustration increases in an already high-stress work environment. This may result in personnel turnover, which can create further stress on remaining workers, leading to a downward spiral on morale, productivity, and project performance in terms of cost overruns and schedule delays.

3

Practical Work Process Management Tips for Success in Project Execution

Introduction

This chapter reviews and discusses successful work process tips that are useful during the execution stage of the project. These processes are by no means new; however, when not properly addressed, they have the potential to have a significant impact on both project cost and schedules during the project execution stage. A project leader or project leadership team should be aware of and responsive to the following common work processes:

1. Do not reinvent the wheel.
2. Ensure proper representations at process hazards analyses (PHAs) and hazards and operability studies (HAZOPS).
3. Use construction and commissioning organizations with consistently high standards of work.
4. Apply simple control systems that work.
5. Ensure ownership and buy-in by all stakeholders.
6. Get big quickly when required and lean in a well-orchestrated approach.
7. Delay process hazard introduction as long as possible.
8. Centralize the work permitting system.
9. Use commissioning and standard operating procedures (SOPs) at all times.
10. Maintain a log of activities and events on critical systems.
11. Have a backup or contingency plan
12. Flush all critical systems before putting into service.
13. Consider human factors in designs.
14. Know when to focus on optimization and efficiency improvements.
15. Ensure an effective document control system exists.
16. Capture and share lessons learned.

Do Not Reinvent the Wheel

In project management, it is important to avoid reinventing the process. When a process works, avoid reinventing it unless it is no longer fit for your purpose. For greenfield projects, there is a tendency for leaders to build from the bottom up with regard to policies, procedures, systems, and processes. Unless there is significant value to be gained from *building new processes and tools,* it is best to apply proven tools and techniques that may already exist within or external to the organization. Leaders should be encouraged to acquire them from effective sharing of information within the organization and from networking with industry peers. As project leaders, we should seek to capture and build on the strengths of tested and proven processes and methods that may already reside within the organization.

For example, there is no additional value in creating new standard operating procedures (SOPs) and codes of practices (COPs) when they may already exist within the organization at another facility or business unit. Generally, policies and procedures may change to meet the site-specific conditions but will change little in substance. Regulatory policies that must be adhered to by the organization are the easiest to apply. These can simply be adopted *as is* consistent with regulatory requirements. Most organizations tend to maintain standards that exceed the minimum requirements of regulations. Resident expertise within the organization can be applied to administer the required training for new employees. Industry-specific policies can vary between competing organizations and can be modeled from industry partners. Often, project leaders may obtain (with approval) regulatory standards adopted by industry peers and modify them to meet the project-specific needs. At most, minor adjustments may be required in making such regulatory standards workable for a new site.

In the project world, there are perceptions that the more information that is available, the better off we are in the long run. As a consequence, project leaders tend to request every bit of information possible from the construction and commissioning organizations in the document turnover process. The turnover process is a simple standard process that should not be redefined. Turnover from construction to commissioning has a fixed set of requirements; similarly, turnover from commissioning to operations has a fixed set of requirements that are specific to the industry, regardless of the player. This process should not be reinvented, and the composition of the turnover package should be changed only marginally unless exceptional value is derived from the proposed changes.

Another good example when this concept is useful is in the way construction work is undertaken and applied within the localized area. For example, in the oil sands of northern Alberta, Canada, modular construction is applied, with modules constructed in the city of Edmonton and shipped by truck through the city of Fort McMurray to the project site directly. This approach to construction has been successfully adopted after years of studies on the comparison between on-site construction relative to this manner of modular construction.

This approach to construction has been so successful that all of the major petroleum industry players in Fort McMurray have adopted the process of modular construction in Edmonton before shipment. Modular construction has the advantages of cost savings and avoids the difficulties associated with labor paucity in the localized area. More important, modular construction drives design engineers to maximize space utilization through more compactly constructed modular units. To consider alternative methods for construction not only would be costly but also would result in long delays from skilled worker shortages in the local environment.

From this discussion, it is clear that significant cost savings and scheduling advantages can be gained from the application of proven and tested procedures, policies, tools, and methods for successful project management. It is important that we avoid adopting the *building up* process if proven methods are available. Time can be more wisely spent by personnel internalizing and applying proven methods versus building new processes and procedures, for which the potential for errors is high and tolerance of errors is low.

Ensure Proper Representations at PHAs and HAZOPS

The PHAs and HAZOPS are essential to the overall success of any industrial project. While PHAs and HAZOPS are generally conducted before the execution phase of the project, honorable mention is required since failures and deficiencies in this phase can lead to major cost and schedule issues when risk and operability concerns are brought into the equation.

It is absolutely essential that the right representation is available during PHAs and HAZOPS for proper identification and mitigation of risks while ensuring the operability of the facility. I strongly recommend the involvement and presence of engineering, process safety management, process engineering, frontline operations expertise, and relevant trades disciplines during PHAs and HAZOPS.

In addition to having the right representation, PHAs and HAZOPS must also occur at the right time in the project cycle. Ideally, such studies must occur in the scoping and design (conceptual design; see Figure 1.1) of the project. At this stage, design engineers must be provided prescriptive guidelines governing process safety and operability of the facility or project. The operations organization has the best opportunity for influencing the quality of the final product at this stage by addressing control systems and safety measures for the facility.

Use Construction and Commissioning Organizations with Consistently High Standards of Work

Imperative to the overall success of any industrial project is the need to use construction and commissioning organizations that demonstrate consistently high performance, behaviors, values, and ethics in work. These organizations generally move from one project to another with a fixed way of doing things. This means these organizations will perform consistently along the continuum from good to bad. They will do things either consistently good or consistently bad from one project to another.

When selecting construction and commissioning organizations, therefore, owners must ensure that these organizations demonstrate consistent and sustained high-performance standards. Adequate screening for incorporating safety in work behaviors, responding to owner's procedures and practices, and strong ethics and values are required to ensure a more productive working relationship among project execution organizations. Often, such organizations are generally available at a higher cost than others. However, higher performance in safety and work quality justifies the additional cost. As the saying goes, "You get what you pay for." If possible, always perform reviews and background checks on both these organizations before they are brought to work on the project work site. It is important to note that once these organizations begin work, should it become necessary to remove them for poor performance, finding suitable replacements becomes a nightmare. In addition, all future poor-quality work encountered will likely be blamed on the prior organizations. Ultimately, the owner and the project suffer from such situations.

Apply Simple Control Systems That Work

During the implementation stage of a project, it is important that simple control systems are adopted and accepted by project leaders. Often, it is best to adopt the "all-or-none" rule to avoid complications from situations that are left to interpretation. Rules must be simple and easy to follow. It is easier to manage a *dry camp* (no alcohol) policy than to administer variations of drinking regarding an alcohol policy. A policy of *three strikes and you are out* regarding speeding in a company vehicle is simple and easily administered as opposed to variations of a speeding policy.

There are tremendous benefits to having simple control systems without compromising personnel safety or the integrity of the project site. Project leaders must bear in mind that when the rules are simple and easily interpreted there are few excuses for failing to comply. More important, this method of administration avoids opportunities for leadership personnel to inadvertently take sides on any situation. Following through on consequences for violation of rules is also important since failure to do so reduces all site rules to a rubber-stamping process that will eventually fail. In addition to being simple, rules must be practical and applicable. For example, a policy requiring clean-shaven facial hair in the avoidance of a breathing hazard may appear unfair and inappropriate. Personnel reaction may result in a destabilization of the work process to unfair and impractical rules.

Personnel access control to the work site is also an important regulation that must be simple and effective. During project execution, there is often an abundance of vendor representatives and service providers visiting the project site with valid reasons for their visits. A system to effectively allow access to the site and to make contact with the right owner representatives on site is important and must be supported. This is best facilitated by having key representatives of the owner's organizations determine who shall have access to the site through a simplified authorization form as shown in Figure 3.1.

Control of personnel to the site grows in importance as the project transitions in ownership to commissioning and operations and process hazards are introduced to the site. This is particularly important in the event of emergency responses or during plantwide evacuations, when accurate personnel counts are required. At this stage, site security takes on increasing prominence, and an accurate personnel count in and out of the access gates is required. All personnel entering the site will be required to receive a site orientation training that defines hazards, emergency response alarms, and the designated muster point in the event of an alarm. In addition, site access may require certain specific regulatory training, such as hydrogen sulfide awareness training when sour gas is present at the facility.

Beyond personnel control is the need to control materials flow to the site and subsequent use and reordering of materials. Simple control systems (e.g., sign in/out control systems from a single point of control) may be required for use of tools and communal equipment. As best as possible, it is important to adopt self-policing mechanisms for control. As site risks increase, however, a more effective policing and control method will be required.

Ensure Ownership and Buy-In by All Stakeholders

Ownership and buy-in by all stakeholders are critical for success in project execution. Stakeholders are both internal and external to the organization and include business partners; business planners; design engineers; communities bordering the project and their leaders; regulatory bodies;

FACILITY / PROJECT: _____

SITE ACCESS FORM

Site Sponsor	Telephone No	Email Address	Affiliation	Safety Orientation Required
Maintenance Manager	Telephone number of Site Sponsor	E-mail address of Site Sponsor	Plant Operations	☒ No ☐ Yes

Details of Visiting Personnel

Name	Site Contact	Service Provided	Services to		Safety Orientation Complete
John Doe	Jim Harris	Vendor Support on compressor systems	☐ Construction ☐ Commissioning ☒ Operations		☒ Yes
Mary Parker	Tom Jones	Review Control systems on centrifuges	☐ Construction ☐ Commissioning ☒ Operations		☒ Yes
			☐ Construction ☐ Commissioning ☐ Operations		☐ Yes
			☐ Construction ☐ Commissioning ☐ Operations		☐ Yes
			☐ Construction ☐ Commissioning ☐ Operations		☐ Yes
			☐ Construction ☐ Commissioning ☐ Operations		☐ Yes
			☐ Construction ☐ Commissioning ☐ Operations		☐ Yes

Date of Arrival	Estimated Time of Arrival	Date of Departure	Estimated Time of Departure

Site Office Accommodation Required

Required	Date	Time Start	Time End	Facility Size	Request Date	Authorized By
☐ Yes ☐ No				☐ 1 to 10 ☐ 1 to 15		

Comments:

ALL VISITORS ARE RESPONSIBLE TO BRING STANDARD PERSONAL PROTECTIVE EQUIPMENT (PPE)

FIGURE 3.1
Site access control form.

construction, commissioning, and operations personnel; vendors and service suppliers; input suppliers; marketers; and end product users. The objectives of the project must be clearly defined and communicated in a manner that allows stakeholder buy-in for maximum contribution to the project. Careful attention to what is being communicated to the target stakeholder must be undertaken to ensure buy-in and commitment by the target stakeholder groups.

Full understanding of the impact of individual contribution to the overall project maintains high motivation and retention of personnel. Ownership resides among all of the stakeholder groups mentioned and slides along a continuum as the project transitions through the various phases of the project cycle. During the execution stage, ownership takes on a complex outlook in which three large stakeholder groups have an opportunity to coexist for a defined period and to see the project transition from blueprints to a producing plant or facility. Construction personnel pride themselves on building it, commissioning personnel pride themselves on testing it, while operations personnel will pride themselves on making it work on a sustained basis. Ownership during execution is strong among all groups, and careful attention to stakeholder interest is necessary.

Careful consideration to managing ownership issues during this period is essential since disastrous outcomes can occur should the efforts and contributions of each group be understated or trivialized. Events such as work stoppages, poor-quality workmanship, poor worker attendance, reduced productivity, and shirking on the job can all be avoided through careful regard to ownership. This is particularly important to the construction group, which will normally be the first stakeholder group to exit the execution phase and whose unplanned departure or absence can leave the process totally stranded.

It is important to point out that not everyone who works on a project site will share the same ownership values. This is particularly true in the case of temporary personnel, whose primary goals most often will be to maximize earnings. Project leaders should seek to distinguish between the behaviors of temporary workers seeking to gain working experience to obtain permanent employment in the future from those who are income maximizers. Ownership, productivity, and overall contributions will be significantly higher as temporary workers seeking permanent employment seek to *overperform* in the eyes of potential employers.

It is important that all stakeholders fully understand and accept the deliverables of the project and trust in the work of the planners and technical experts who had the capability to capture seed ideas and translate them into plans and designs to deliver finite results. In the absence of this understanding and acceptance, motivation among all stakeholders is likely to be low, and there will be a failure to retain and attract quality personnel to the process. Overall, the project is likely to suffer cost overruns and delays.

Get Big Quickly When Required and Lean in a Well-Orchestrated Approach

Growth of the project execution organization is an important variable in the overall success of a project. Timely introduction of the commissioning organization and the operations organization helps to keep the project on the proposed budget and scheduling path. Figure 3.2 shows the direction of growth of the organization. While direction of growth of the organization is important during the execution stage, the manpower loading at the project site is of greater importance. In Figure 3.2 (a typical manpower loading plan), we see a near-exponential growth rate for the construction organization. Once construction has been initiated, there is a rapid and sustained growth of both the commissioning and the operations organization up to peak manpower loading. When personnel from the operations and commissioning organizations are added to the manpower loading, an even higher peak manpower load is achieved in a very short period.

From my perspective, project execution is best accomplished when using a commissioning organization that eventually transitions into, or forms part of, the permanent operations organization. However, this is not always easy to accomplish.

By the time the commissioning organization is fully staffed, the operations organization is in advanced stages of development. The near-exponential growth as shown on the curves highlight the emphasis on getting big fast. For the operations organization, getting big quickly is a necessary requirement because full understanding of the various pieces of equipment is best accomplished as the pieces of equipment are being constructed, assembled, and installed by the construction organization. Similarly, it is best to inspect

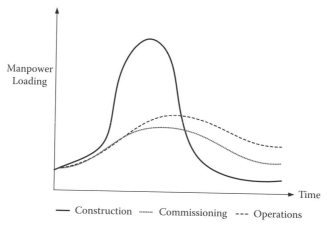

FIGURE 3.2
Typical manpower loading profile.

FIGURE 3.3
Finished storage tank ready for service. (©Suncor Energy Inc. With permission.)

internal components of vessels and equipment during construction and before turnover of equipment and assets to the commissioning organization.

Figure 3.3 shows the profile of a completed tank that is ready for turnover from the construction organization to the operations organization. When completed, there is not very much to see to understand how the tank operates. Figure 3.4, on the other hand, shows some of the internal components of the tank. Shown in the top panel is the distribution system of the incoming fluids into the tank. Open-ended pipes are provided to minimize turbulence within the tank, and flow is distributed uniformly throughout the tank. Shown in the bottom panel is the bottom of the tank, which shows the following components:

1. Sacrificial corrosion anodes, which will be preferentially corroded, are found around the perimeter of the base of the tank.

2. A chemical injection and treatment line is for shock treatments in the event of process upsets.

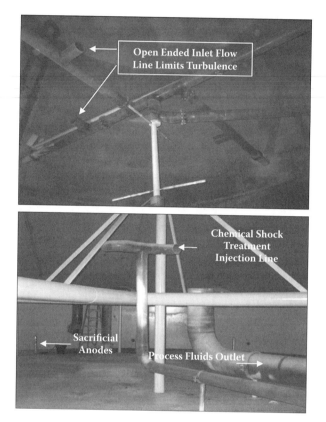

Open Ended Inlet Flow Line Limits Turbulence

Chemical Shock Treatment Injection Line

Sacrificial Anodes

Process Fluids Outlet

FIGURE 3.4
Internal components of finished storage tank. (© Suncor Energy Inc. With permission.)

3. The process fluids outlet line from the tank is covered by a cone-shaped structure to minimize turbulence.

The first group to leave the project execution stage is the construction organization on completion of major construction work. A small construction staff may be left behind to complete all outstanding deficiencies and noncritical work not completed during the peak construction periods. Generally, the construction organization is by far much larger than any of the other organizations.

Getting big fast is not always easy for a number of reasons, such as market availability, skill level of available workers, state of the economy, labor mobility, and employment levels. For the construction organization, this is particularly important since the amount of construction activities taking place in a region may affect availability and quality of the construction workforce.

It is important, however, that critical personnel and discipline leaders be hired early enough to support the growth of the various organizations, particularly the operations organization. The recruitment and selection process can be very long. For this reason, the hiring process should begin early enough to ensure adequate staff is available at the right time in the execution stage. A rule of thumb is that the leadership team (construction, commissioning, and operations leaders) should be hired well before the project execution begins so that they may participate in the planning work prior to the start of construction.

These leaders may then have opportunities to assess the work required by their respective organizations and based on their experience levels will assess the manpower loading and timeliness of their hiring plans to meet the needs of the project and to avoid schedule delays. In addition, the leadership team will have the opportunity to assess their hiring practice to determine both availability and accessibility of skilled personnel and to determine what is required to attract these personnel.

It is not uncommon in most instances for potential employees to provide 2 weeks of notice to an employer prior to leaving an organization. This must be factored into the recruitment and selection process when growing the various organizations. In some regions, growing demand for skilled personnel to staff the three organizations of the execution stage can be an extremely difficult proposition. Attracting and retaining skilled personnel in this area may require creative means for staffing these organizations, which may include:

1. Higher wage rates and salaries for comparative skills in other industries
2. Relocation assistance and family retraining and job search incentive plans
3. Signing bonuses that may contain *locked-in* clauses for 2–3 years of employment
4. Retention bonuses to prevent employees from moving from one organization to another in regional markets
5. Career growth and development opportunities for skilled and talented performers
6. Housing and accommodation incentives in some regional markets

Leadership must understand the dynamics of the local market and take appropriate steps to ensure adequate personnel are available precisely when they are required. When not clearly understood, schedule and cost implications for the project execution stage are highly probable. The construction organization generally operates as a complete unit, and once a contract is awarded, there may be the translocation of a complete organization to the work site with all amenities provided to sustain this cohesive group. Both the commissioning and operations organizations have to be built up since these types of cohesive groups are not readily available in the market. Project leaders must understand that both the commissioning and the operations organization have huge tasks ahead to ensure continued and timely project execution. Some of the major tasks include those identified in Table 3.1.

Clearly, personnel can become overloaded quickly if the organization does not develop at the required rate to accomplish the tasks defined in Table 3.1. Hiring practices and skill capabilities of workers can significantly influence the workloads of individual commissioning and operations personnel. The presence of inadequate or unskilled personnel for an extended duration can lead to situations of work-related stress, burnout, and increasing safety infractions and incidents. In addition, leaders may experience increasing difficulties in scheduling vacation and breaks for workers. This can lead to vacation carryover to the subsequent year, which may result in a logistical nightmare for the operating organization in administering vacation and breaks in the future.

To guard against these undesirable social management constraints and issues, it is important that the organization keeps abreast of project require-ments and encourages the construction, commissioning, and operations organization to grow quickly. Project leaders must understand the impact of market-influencing factors and mitigate against them. Failure to do so can result in cost escalation from human-related errors and incidents, an over-worked staff, poor work morale and productivity, lost knowledge and exper-tise from employee turnover, and schedule delays.

As the project approaches completion, a managed approach to reducing the size of the workforce is essential. If the workforce is reduced too rapidly, the project may lose momentum toward the completion schedule. On the other hand, if too many workers are retained, the potential for reduced worker productivity increases as more workers are available to perform too few jobs. Project leaders must therefore strike the right balance between retaining per-sonnel to manage the schedule and reducing the workforce.

Delay Process Hazard Introduction as Long as Possible

The introduction of process hazards into the work site has a profound impact on the work safety practices to be applied in the workplace. Generally, when the construction organization owns the workplace and process hazards are not yet introduced, the work pace and worker productivity are high, and risks are low and associated primarily with construction-type incidents.

As the commissioning organization becomes active and systems are ener-gized with the introduction of process fluids, operating pressures, and elec-trical power, work practices change dramatically in response to the newly introduced hazards. Work permitting is now adjusted to address these newly

TABLE 3.1

Sample Tasks of the Commissioning and Operations Organizations

Commissioning	Operations
• Develop commissioning procedures for critical equipment and systems.	• Develop standard operating procedures (SOPs) for all equipment.
• Arrange and schedule vendor testing of equipment and conduct equipment-specific training.	• Develop and execute site-specific policies based on prime contractor's obligations.
• Identify and schedule all temporary requirements to properly execute commissioning of equipment. This may include temporary piping, pumps, and compressors.	• Develop training manuals for training new recruits and operations staff. Conduct, execute, and track training requirements in a visible framework to ensure all personnel are ready to assume the task at hand.
• Participate in punchlisting exercises to successfully receive systems from the construction organization.	• Work with the commissioning organization to ensure punchlisting is completed as required to successfully receive systems from the commissioning organization.
• Review and compile documentation required by the operations organization in order to maintain continuous operations of the plant.	• Ensure that the safety culture framework begins to take hold and define the requirements for the future of the organization.
• Develop a process to ensure equipment received that does not meet contractual obligations is addressed in a timely fashion so the process is not hindered.	• Manage the readiness process for start-up and takeover of plant on a system-by-system basis from the commissioning organization.
• Develop a process for addressing design and process deficiencies.	• Define the plant into operating areas, risk rank the operating areas, assess skills capabilities of operating personnel, and assign personnel based on risk and skills capabilities.
• Provide ongoing support to the operations organization as required.	• Secure and administer operating service contracts.
• Compile and provide turnover documentation consistent with prior agreements.	• Develop the management framework for maintenance of equipment and system received from the commissioning organization.
• Complete outstanding deficiencies before full turnover to the operations organization.	• Develop the framework for long-term maintenance of the plant as a whole. Most often, a computerized maintenance management system (CMMS) will be developed for longer-term application.
	• Critically rank equipment for warehousing spares and consumables.

introduced hazards, and as such personnel must now proceed much more cautiously based on the introduced hazards. The operations and commissioning organizations have a greater role in ensuring the safety of personnel and the protection of equipment, machinery, assets, and the environment.

This phenomenon is much more obvious on shared systems, for which substantially completed systems are turned over to the commissioning organization, which in turn may have completed its testing and turned it over to the operations organization. Live system identification and appropriate isolation mechanisms for joint ownership are required for defining boundaries. Boundary demarcations by valve isolations will require that valves be chained in the closed position with locks placed on the chained valves from representatives of the construction, commissioning, and operations organizations. Important boundary demarcations may include fuel and natural gas valves, steam headers, and toxic and high-pressure process fluid boundaries.

Process hazards introduction or livening up of systems may include, but is not limited to, electrical power introduction to the facility and its distribution centers, introduction of fuel gas into the facility, and introduction of process fluids and operating pressures in any system. The introduction of such hazards introduces the adoption of zero energy isolation practices to allow personnel to work in a safe environment. Zero energy isolation requires that systems being worked on must be free of process fluids, fire and explosion hazards have been removed or mitigated, electrical and power sources have been isolated, operating pressures have been reduced to ambient, and all potential hazards from the workspace have been controlled, isolated, or mitigated. In such situations, system ownership and control is of utmost importance, and joint work on such systems must be carefully managed to prevent incidents and injuries. System owners must understand the dangers associated with their respective systems and must advise all personnel working on the system of the inherent dangers and the mitigating actions applied to manage the risks associated with each system. This is most often done through mass communication, the permitting process, and central control of the permitting.

Despite the assurances provided by the work permit process, the work pace is reduced particularly by the construction organization as it adjusts to the newly introduced hazards. In the process environment, there is a huge dependence on the four major senses of smell, hearing, sight, and touch. At this stage, construction personnel become more sensitive to these senses and respond with great alarm to any changes in them. There are heightened reactions and responses to the fear of the unknown by construction personnel and new and inexperienced employees. Loud noises that were previously accepted as part of the norm can generally result in work stoppage. Strange and unknown odors can trigger adverse reactions from workers and can result in work stoppages. Unexplained smoke, dust plumes, leaks, or unfamiliar sounds and smells can also lead to work stoppages.

At this stage, there is a critical need to centralize the work permitting process and communicate all activities to work group leaders and forepersons.

Project leaders cannot overemphasize the need to communicate to all stakeholders any changes that have an impact on the workplace. While this approach is essentially a mitigation plan to deal with process hazard introductions, true comfort and sustained high productivity from the construction organization comes from delaying the introduction of process hazards as long as possible. Ultimately, the decision to delay or introduce process hazards lies in the hands of the project leadership team and their ability to manage the work process and to mitigate against the introduction of risks associated with process hazard introduction while maintaining high worker productivity. With the introduction of a new process hazard into the workplace, there is a critical need to communicate this event. Project leaders must use the communication methods discussed to inform all personnel of the new hazard. Town hall meetings, e-mails, and toolbox meetings are essential in communicating the new hazard introduction. Figure 2.7 provides a sample notice for the introduction of a new hazard to the workplace.

Centralize the Work Permitting System

What does centralizing a permitting system entail? A *centralized permitting system* refers to joint decision making regarding work to be performed on various systems during this phase of the project execution stage in a multiownership scenario. During the execution stage of a project, work permitting should originate from a central location regardless of the site ownership. The work permitting office should be experienced in the hazards associated with a project environment, and the permit coordinator must possess excellent communication, influencing, and risk identification and management skills. Early in the execution stage, work permits may be administered and managed by the independent construction, commissioning, or operations organizations. However, as work occurs within the three organizations, a centralized permitting system is an absolute requirement for maintaining safety and high work efficiency.

It is important to note that joint decision making is not necessary until more than one system owner coexists on the same site. Centralizing the permitting system with a single point of control has several advantages, including improved site safety, easier workflow control and management, and more effective control of concurrent work on multiowner systems.

Improved Site Safety

During the execution stage of a project, the safety of all personnel is foremost in the decision-making process. Incidents, accidents, injuries, and fatalities all have a huge impact on worker morale, overall productivity, and project cost. A well-managed and centralized work permitting system assists in reducing the possibility of incidents, accidents, injuries, and fatalities. Project leaders must be careful to point out that permitting in itself does not make the workplace any safer. Rather, the work permitting process is designed to highlight the risks associated with assigned work and to ensure system owners have reduced risks to as low as reasonably practicable. In addition, the work permitting process highlights the personal protective equipment (PPE) required by personnel who will be performing the work.

Of key importance is the verification of work preparation activities for rendering the task safe and the adherence to specified conditions determined by the operating authority while the performing authority is doing the work. A skillful permitting coordinator will ensure the conditions defined in the work permit are clearly understood by the performing authority and that all dangers and mitigating factors are properly reviewed before the task can begin.

Easier Workflow Control

During the execution stage of the project, it is necessary for project leaders to adopt and apply simple work control systems that are effective in progressing high-productivity work in a safe manner. A centralized work permitting system allows for all personnel to begin work at a specific time and in a controlled fashion while minimizing bottlenecks, avoiding confusion, and permitting work to be done safely. On a project site with construction, commissioning, and operations personnel working together, there will naturally be a multiplicity of trade groups and different work activities taking place at the same time. When all personnel begin work at the same time, the potential for bottlenecking and confusion is high if work permits are demanded at the same time by all work groups or organizational representatives. A centralized work permitting process helps to streamline work among all groups by adjusting the start times of each group.

It is important to note that for this process to work well, organizational leaders and the permit coordinator must align start times to ensure the smooth working of the permitting system. Leaders from the three organizations must be present to validate and agree with permitted work. This process is best demonstrated in Figure 3.5.

As shown in the example provided in Figure 3.5, construction trade leads and forepersons begin their day at 0530 hours, a half-hour earlier than their work groups. This is done so that their work permits can be validated and approved to begin work at 0600 hours. Similarly, commissioning leads start their day a half-hour earlier at 0600 hours to have their permits approved and to commence work at 0630 hours. Finally, operations leads start at 0630 hours to commence work at 0700 hours.

Once the bottlenecking period (which normally occurs at the start of the day) is over, work permitting can continue for all groups from the centralized center by the permit coordinator only. Input from the various system owners will be solicited by the permit coordinator and the permit requesting

FIGURE 3.5
Workflow management through a centralized work permitting process.

party based on the risk associated with any particular job. Input from organization leaders can be requested as required. This approach to permitting continues until there is a substantially full transition of all systems to the operations organization.

From this point on, the operations organization assumes full responsibility for the site, the safety of all personnel and equipment, workflow, and continuous operations. Success of a centralized work permitting process as described depends on an effective work planning process in which work is planned at the close of the previous day and work preparation consistent with the requirements of the work permit is accomplished by a night shift team.

Effective Control of Concurrent Work on Multiowner Systems

During the execution stage of the project cycle, at some point in time the construction, commissioning, and operations organizations will perform work in the same work area or on the same system at the same time. Furthermore, the work conducted by one group is likely to have an impact on the work being conducted by another group, creating an unsafe condition for some or all of those who are working in the area.

The centralized work permit system allows the permit coordinator, with input from the system owners, to schedule the work of the impacting group for a period when the impacted groups are away from the specific area or system (e.g., during breaks or lunch periods). In this way, concurrent work can be undertaken without loss of continuity. On the other hand, it may be necessary to remove and reallocate other work groups to enable critical work to proceed.

A good example of concurrent work can be the overhead lifting of equipment to its permanent site while other work is being undertaken at lower levels in the same area. Depending on the duration of the exercise, the work of other groups can be rescheduled, or the groups themselves may be reassigned to other areas where they can work safely and without interruption. To protect the safety of workers, such lifting can be done during lunch breaks, by the work group after all workers have left for the day, or by removing other workers from the site to allow the lifting to proceed.

Sometimes, it may be necessary to seek the right opportunity when work can be performed during the available window of opportunity. A good example of this can be a lifting exercise that can only be accomplished when the lifting equipment is available unless additional lifting equipment is brought to the project site. With a centralized permit center, the permit coordinator can seize the opportunity when the lifting equipment is available for performing the job while minimizing work interruptions.

From this discussion, without exception, a centralized work permitting system adds significant advantages to work continuity and sustained high productivity. For the project execution stage of the cycle, a centralized work permitting system is a necessity for continued high productivity and a safe work environment.

Use Commissioning and Standard Operating Procedures (SOPs) at All Times

Well-constructed Commissioning and SOPs are designed to ensure equipment and machinery are operated properly and to avoid unintended damage to the equipment and machinery. The use of procedures also helps inexperienced workers start up and operate equipment safely. The sequential steps and tasks defined in operating procedures provide a necessary road map to workers who may never have performed such tasks. Procedures also enable them to do so safely and without damaging equipment or machinery or hurting themselves.

In the event of an incident when commissioning and starting up equipment or machinery or a system, investigators will gravitate to the commissioning or operating procedures to determine whether workers followed the instructions defined in the procedure. Failure to follow procedures will ultimately place the responsibility for equipment damage on the laps of the organization that was working at the time of equipment failure.

The use of commissioning and operating procedures must be encouraged by all personnel. Leaders must encourage and enforce the use of procedures when starting up or putting equipment into operating service. During equipment start-up, printed copies of the procedure must be taken to the field by operating personnel. Personnel must follow each step defined in the procedure and must indicate so by affixing a signature next to each step once it is completed. If a step is identified as obviously wrong and can potentially damage the equipment or hurt personnel, work must be stopped, and the procedure must be revised with the approval of the appropriate level of authority. Control over modifying commissioning and operating procedures must be maintained to preserve the integrity of the procedure. Once the equipment or system has been commissioned and started, all procedures used in the process must be reviewed for improvement opportunities. Procedures used in the commissioning and start-up of equipment and machinery and systems should be retained on file until the equipment and systems are deemed to be operating satisfactorily.

Maintain a Log of Activities and Events on Critical Systems

A log of activities and events on critical tasks is an important requirement during the commissioning and operations phases of the execution stage of a project for validating warranty and liability claims. Many problems and concerns can arise with newly installed equipment and machinery. Commissioning and operations personnel must be prepared to establish and provide historical information databases on activities to ensure costs associated with repairs or damage related to equipment are properly assigned and recovered.

On vendor-supplied equipment and machinery, every effort must be made to ensure vendor representatives are present during commissioning and start-up of critical equipment and machinery. Such activities must be properly scheduled in the plan to allow for vendor availability. Vendor presence is important so that unintended failures and malfunctioning of vendor-supplied equipment and machinery can be witnessed. Furthermore, equipment can be started up in accordance with vendor specifications and under the supervision of the vendor. If vendors are unable to be present during critical equipment start-up, procedures provided by the vendor must be followed. A documented log of events must be maintained in all cases to establish due diligence in the event of warranty and liability claims from damage and failure.

For joint operations in which the potential for liabilities exists, it is paramount that project leaders ensure that a well-documented historical trail is retained on issues for which either organization may be liable. For sensitive issues, it may be necessary that a single individual be assigned the role of maintaining daily log details. The importance of maintaining a running daily log of activities complete with dates, activities, personnel, and observations cannot be underestimated since large project costs may be associated with warranty, repair, and liability issues.

Have a Backup or Contingency Plan

Having a backup or contingency plan is an absolute necessity for both the commissioning and operations organizations during the execution stage of a project. Reality has shown in many cases that in spite of the many hours involved in detailed planning and coordinating, many activities will not work exactly as planned. As a consequence, leaders will be required to resort to a backup or contingency plan. The backup or contingency plan can be initiated in response to readiness criteria for people, process, and system readiness. Backup or contingency plans can range from the simplistic reallocation of resources to the elaborate redesign of systems.

Some criteria for each of the readiness variables that may require leaders to evoke the backup or contingency plan are discussed next.

People readiness delay criteria:

1. Operating procedures not ready and available
2. Operations/commissioning personnel not fully trained and competency assessed
3. Inadequate personnel availability to sustain process

Process readiness delay criteria:

1. Adequate monitoring processes not yet in place
2. Processes not yet in place to address unacceptable operating conditions and product quality
3. Adequate accounting processes not yet in place

System readiness delay criteria:

1. Delays in equipment setup or availability or components for system operation
2. Immediate failure of critical equipment during commissioning
3. Inappropriate design determined at the time of commissioning

For each of the delay conditions identified, a backup or contingency plan is required to proceed with the project execution. In some instances, the fallback position can be the decision to halt the process until the delay condition or deficiency is resolved. This is particularly important when the risk of injury to personnel is high or the cost associated with damage to equipment may be high. Sometimes, the time associated with replacing the damaged equipment may also influence the decision to proceed; often, this decision will be determined by the objectives of the corporate organization as a whole.

As discussed at the beginning of Chapter 1, *good news* is associated with favorable market responses. In most instances, schedule delivery appears to be a more attractive option than cost overruns, particularly if the cost overruns are not excessive. A conscious decision may be taken by the project leader to incur the additional cost, if any, associated with the next-best alternative so that stakeholders can be heralded with the good news that the project continues to be on schedule.

If unacceptable risks arise in the process, such risks can be reduced to manageable and acceptable levels with sufficient engineered or administrative controls to ensure the safety of personnel or to minimize damage to equipment. When backup plans must be developed on the fly to address a newly encountered situation, the skill capabilities of personnel will be tested to identify and execute potential solutions. It is important, therefore, that the leadership team, particularly the commissioning and operations leaders, have experienced and capable resources within easy access to assist in developing and assessing possible alternative solutions to technical problems that may arise.

Generally, backup or contingency plans are rooted in *what-if* studies and should be kept current in the event it is necessary to use them. The importance of backup or contingency plans cannot be understated during the project execution stage. Proper planning is required not only for the project execution stage of the cycle but also for longer-term and ongoing operations. Indeed, contingency plans must also consider the impact of reduced productivity resulting from extreme weather conditions such as winter in temperate countries and high temperatures in arid and desert regions. The absence of well-defined backup or contingency plans in identified situations can make the difference between delivering a project on schedule and within budget and being late and over budget.

Flush All Critical Systems before Putting into Service

Often, it may be possible to commission and start up a system without the need to flush or steam clean a system. For critical systems, flushing and cleaning the system to remove debris, contaminants, mill scales, and unwanted materials that may affect operating equipment in the system must be done. In addition to a physical inspection before a system is closed or buttoned up, critical systems must be water or acid washed; in some cases, a steam blow may be required to ensure the system is cleaned properly before being placed into service.

In the absence of proper cleaning, metal debris and other contaminants may damage or have an adverse impact on the proper functionality of downstream equipment. For example, in a high-pressure boiler feed water system, foreign matter in the piping or storage equipment may lead to damaged impellers, seals, and valves in the system. Even more important, such debris and contaminants may lead to a complete facility shutdown when debris becomes lodged in the body of a main valve and the valve can no longer be closed properly, leading to process upsets and personnel and equipment hazards.

The use of in-line suction screens in pumps and prefilters to critical equipment is crucial during this stage. Failure to include them can be disastrous to equipment and can result in significant delays during project execution if replacement equipment is not available in stock or a long lead time is required for reordering. Furthermore, the cost-related impact from construction-related delays can be phenomenal. Figure 3.6 demonstrates the benefits of suction screens to pumps (commonly called *witches' hats*) in collecting debris before it can enter a pump casing to cause damage.

A steam blow is an effective method for removing foreign materials from long sections of unrestricted pipes. Low- or medium-pressure steam may be introduced at one section of a long section of pipe, and with appropriate warming up of sections of the piping, a controlled steam blow is possible. Steam blows are effective in removing debris inadvertently left in process piping. Steam blows are also effective in removing mill scales from piping.

Once the critical system has been thoroughly flushed and there is confidence that all debris have been removed, the in-line suction screen can be removed. In some systems, the screen may be a manufacturer requirement, in which case it must remain in service. Frequently, however, in viscous fluid services, the suction screens can lead to an unacceptable drop in pressure across the filter and must be removed to prevent damage to the pump.

Figure 3.6 demonstrates some of the debris collected when a suction screen is used to collect foreign materials before they have an impact on critical equipment. Figures 3.7 to 3.9 demonstrate the impact of failure to remove

(a)

(b)

FIGURE 3.6
In-line suction screen protections for flushing critical systems. (a) Suction screen bolted in-line between two flanges on the inlet suction line to a pump. The size of the sieves must be such that it does not restrict flow to the pump and result on restricted flows and potential damage to the pump from cavitations. (b) The piece of metal debris shown in this suction screen can cause tremendous damage to a pump rotating at high speeds. This easily removable suction screen, once filled with debris, will result in reduced flow to the pump and must be removed for cleaning and replaced once finished. (© Suncor Energy Inc. With permission.)

(a)   (b)

FIGURE 3.7
Impact of foreign materials in a system on the valve trim of a control valve. (a) Trim of a control valve plugged from foreign materials in the process fluids. (b) Same valve trim after being cleaned. (© Spartan Controls. With permission.)

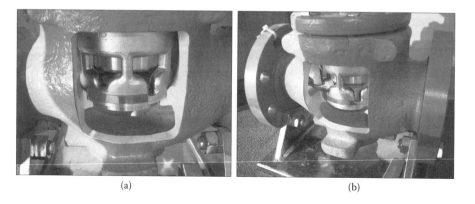

(a)   (b)

FIGURE 3.8
Globe valve with section removed to show impact of debris in valve seat and its inability to close. (a) Cut out section of a normal globe valve seating and closing properly. (b) Same valve showing a piece of metal debris caught in the valve seat and preventing the valve from seating and closing properly. (© Spartan Controls. With permission.)

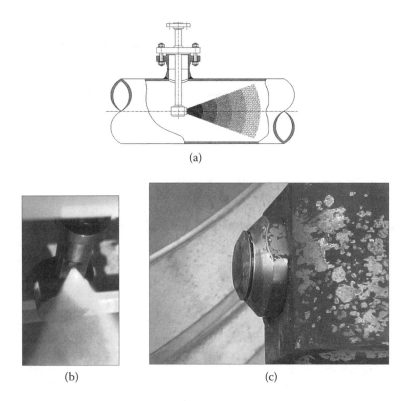

(a)

(b)   (c)

FIGURE 3.9
Distinction between a normally functioning desuperheater system and a system affected by debris. (a) Schematic of a normally functioning desuperheater system. (b) Normally functioning desuperheater system. (c) Debris caught in the desuperheater valve mechanism, resulting in abnormal sprays and a poorly functioning desuperheating system. (© Spartan Controls. With permission.)

foreign materials from a system when in operation. Figure 3.7 demonstrates the impact of failing to include screens or filters on a system upstream of a control valve in the system. Figure 3.8 shows metal debris caught in the seat of a valve, preventing it from closing properly. Figure 3.9 shows the impact of debris on the normal functioning of a system.

Consider Human Factors in Designs

While human factors consideration may be outside the scope of project execution, it is imperative that project leaders ensure that human factors are considered in the design stage of a project. Human factor considerations during design of a facility are critical for success during and after the execution stage of the project. Human factor considerations address the human/machinery interface in the commissioning and operations of the facility. Poorly designed systems, tight spaces, and difficult-to-service systems can result in frustration among operating personnel. More important, tight process designs leave little room for process and worker maneuverability during project start-up.

Make Facility Operator Friendly

A system that is designed to avoid fluids contaminating the ground must be equipped with an adequate system for draining and collecting fluids so that contamination can be avoided. Equipment must be have adequate drainage to properly prepare it for preventive maintenance. For example, a pump with a volute capacity of 100 liters must have a drainage sump of at least 110 liters to properly accommodate the process fluids.

Similarly, valves that are located at heights must be accessible either via a chain system or via easily accessible platforms or permanent ladders. In the absence of an operator-friendly workspace, personnel will consider every available opportunity to leave the job to find a work environment that provides a friendlier human interface. Figure 3.10 demonstrates the use of platforms, ladders, and chains for accessing valves that must be operated on a frequent basis.

Avoid Tight Designs

Tight process designs can result in equipment damage and the need for temporary facilities to accommodate process operations when they occur. Tight process designs can increase operator stress and are generally not recommended in high-risk process systems. An example of a tight process design may be the need to blowdown a boiler at a rate of 400 M^3 of hot boiler feed water for a period of 10 hours during a testing process and maximum storage capacity of 4,000 M^3.

While some evaporation may be expected, there is little room for errors or unforeseen circumstances during testing. Any delays in completing necessary work during the blowdown process will lead to blowdown water exceeding the storage capacity. Such a situation may result in a shutdown of the work being done until the storage facility can be emptied to accommodate the required blowdown water. When the same testing process must be completed within a continuous run period of 10 hours and storage capacity is exceeded, temporary pumps, piping, and storage tanks may be required to ensure the work undertaken is completed.

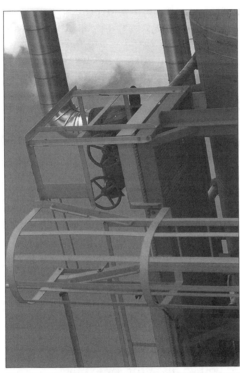

(b)

(a)

FIGURE 3.10
Human factor consideration in design of a facility. (a) Chains attached to valve wheels allow for easy access and operations at ground level and improve the worker/equipment interface. (b) Ladders and platforms allow for easy access and operations of valves, leading to improved worker/equipment interface. (© Suncor Energy Inc. With permission.)

Know When to Focus on Optimization
and Efficiency Improvements

Knowing when to focus on optimization and efficiency improvements is critical for any project manager. Often, when a project is turned over to the operations organization, leaders of the organization are of the view that the project should produce at design capacities quickly. For simple projects for which the technology is proven and well established and all the operating variables are known and tested, this is a reasonable expectation.

On the other hand, when technology is new and the project is among early adopters, organizational leadership expectations may be much higher than what the project may actually be capable of producing. In most instances, it is often difficult to replicate similar operating conditions experienced in a pilot study with a large commercial project designed and built based on the pilot.

For example, in steam-assisted gravity drainage (SAGD) technology, which is relatively new, operating conditions such as temperatures, pressures, chemistry, and residence times that are easy to maintain in pilot studies are extremely difficult to replicate in commercial operations. As a consequence, as far as new technology is concerned, the operations organization may have the responsibility for developing an entirely new set of operating parameters for operating at a commercial level. Essentially, the project has now taken on the appearance of a pilot project at a larger scale, and determination of these new operating parameters may take years based on new operating conditions.

While the new operating parameters are being developed, there may be a tendency by owners of the project to superimpose optimization and efficiency measures. Pursuing both goals at the same time can be conflicting at times, resulting in more delays and cost excursions. For example, as far as water chemistry is concerned, it may become necessary to inject chemicals at a rate that is significantly higher than that defined in design to ensure adequate water treatment. It may be necessary to inject chemicals at two or three times design rates to ensure quality separation of water from oil.

Once the separation process is working properly, chemical injection rates can be reduced to determine the optimum water chemistry requirements for acceptable water and oil separation. If the facility is operating at half of its designed throughput, as production increases, operating conditions may change further, thereby requiring a similar process for determining the optimal chemical injection regimes to promote adequate separation at the new production levels.

Pursuing parallel goals of production increases with cost control can result in conflicting priorities when chemical usage is concerned. If the project is not perceived by owners as a new pilot project at a larger scale of production, conflicts between optimization and operating conditions may arise, resulting in cost increases and further delays in getting to designed production levels.

The focus therefore should be on getting the project to deliver on production parameters while new learning occurs. Once this learning has occurred and the new operating parameters have been defined, optimization can occur to improve operating, technical, and commercial efficiencies.

Ensure an Effective Document Control System Exists

During project execution, there is a critical need for ensuring effective document control after documentation packages have been turned over to the operations organization. While all three organizations co-inhabit the site as final construction, commissioning, and deficiencies are completed, the documentation used to guide all work groups may be that turned over to the operations organization. Careful management of the documentation is required since these documents will be required to perform maintenance work on equipment, to manage change in design and process, and for insurance purposes in the event of an unplanned occurrence.

A document control and management system that requires sign out and in for documents and systems packages is essential for ensuring control. As far as critical documentation is concerned, more stringent control in use is required. Ideally, two or three copies of documentation are required. Documentation can be bulky, and storage and management may require a designated area complete with photocopying services and an administrator to manage compliance in usage.

Ideally, I recommend retention of an electronic version for all documentation. Storage and retrieval of electronic documentation can be managed much more easily. The downside of an electronic document control system is the initial requirements for an electronic version. Often, supplier information and equipment check sheets and marked up (redlined piping and instrumentation drawings) and turned over from one organization to another in hard copy versions. Conversion to electronic versions via scanning and other methods can be costly and labor and time intensive. Furthermore, once electronic versions are developed, an effective mechanism for version control is essential to ensure up-to-date information is available for end users. In most cases during project execution, document controls tend to be an area of weakness.

Capture and Share Lessons Learned

During project execution, there are many project activities that will be repeated over time. Learning captured from previous similar activities makes future attempts much easier. A key competency of a successful project leader is the ability to recognize, capture, and share lessons learned and knowledge (successes and failures) that are transferable for application in future projects.

It is important that project leaders apply processes that work well and avoid those that are difficult to administer. Understanding root causes of failures is also important to avoid repeats. Project leaders must leverage strengths in areas of expertise and seek to improve performance in areas in which weaknesses may exist. Generally, areas of strengths and weaknesses are best demonstrated in captured learning.

Learning from things that were done well and those that were done poorly should be captured for future use. Learning must be captured from high-impact areas that have the potential to influence both the cost and the schedule of the project. Contract administration, procurement systems, the recruitment and selection process, training and development processes, safety management protocol, safe work management and permitting, risk management processes, and turnover processes are all areas for capturing learning.

Project leaders should be encouraged to capture learning from every level of the process and encourage leaders to *think outside the box* in the execution stage. Capturing learning requires leaders and workers of the three key phases of execution—construction, commissioning, and operations—to work with their teams to ensure learning in key areas of successes and failures are captured, standardized, and shared across the owner's organization to promote continuous improvements across the owner's business.

Learning can be captured through simplistic processes such as suggestion boxes, all-inclusive participation via e-mail, or *read/write access* by all stakeholders to an electronic folder aimed at capturing the learning. At the other extreme, a more elaborate structure with adequate screening gates and approval board can be applied. For the ongoing project, however, the simplistic method may be most applicable for immediate implementation. Capturing learning starts at the beginning of the project execution stage and stops after the project has been deemed to be an operational facility that is meeting its designed targets.

It can be difficult to tabulate and assess all suggestions on an ongoing basis as the project progresses. However, it is important that a periodic review of the suggestions received is conducted so that immediate improvement suggestions can be introduced to the process. More important, implementation of good suggestions sends strong and powerful messages to all participants that their input is important, that project leaders and the organization as a whole are serious about making improvements to the process, and that the involvement and contributions of all workers are valuable. Early adoptions of cost-saving suggestions and those that can significantly influence the schedule also send a strong message about leadership's intentions on cost and schedule control. Such behaviors promote the behaviors among workers required for capturing learning.

Conversely, weak and tardy responses to cost-saving suggestions send the message to the average worker that leadership does not care about cost control, so why should the worker? Likewise, suggestions that have schedule-influencing impact must be evaluated and acted on in a timely fashion to avoid creating a lack of interest and involvement among workers and to capture the value they bring.

It is important to note that not all suggestions can be acted on immediately. Sometimes, the learning is identified well after the project is implemented and should be captured in the event that similar projects are undertaken in the future. Often, the adoption of suggestions and enacting change in midstride is a difficult proposal that must be deferred and can be captured for future execution.

Incentives and recognition programs help to promote the capturing and sharing of learning. Personnel can be motivated to capture and bring forward learning by offering incentives or appropriate recognition for such learning. Methods of recognition may vary from the simplistic mention at a morning toolbox meeting as workers discuss the day's activities to mention in the corporate communication systems. Major contributions can also be recognized by more tangible rewards.

The benefits of capturing and sharing lessons learned cannot be underestimated. A simple, effective mechanism for collecting, standardizing, and sharing them must be available. If the process becomes too elaborate and difficult to use, people will veer away from providing learning. Key learning during project management must be captured at every stage of the project, since both organizational and industry improvements can be achieved through this shared learning.

4

The Readiness Processes: An Overview

Introduction

During the execution stage of a project, an astute project leader will determine milestone events to define execution progress to stakeholders. The movement from one milestone to another is reflected by the perceived readiness at that milestone to declare success and to move to the next milestone event or activity. The way in which readiness is determined is dependent on a weighted average process that focuses on the individual readiness of the key components of people, processes, and systems (PP&S) at a particular milestone. The weighted average readiness process is a practical tool for assessing the readiness of the main components of a project as project execution proceeds from one milestone to the other. This chapter examines how the readiness process works and the method used for developing the readiness framework during project execution.

The readiness process provides a simple quantitative assessment of the overall readiness of a project based on the readiness of the individual components of PP&S. Overall milestone readiness is presented in a quantitative manner and expressed as a percentage. Milestone readiness also reflects the weighted average readiness of the individual components based on a predetermined weight assigned to each component at the specific milestone of the project. It is important to point out that the summation of weights assigned to PP&S is equal 1.0 for each milestone. Readiness at each milestone is the sum readiness of PP&S readiness for that milestone and cannot exceed 100%.

The readiness of each component is determined by the readiness of each criterion that makes up the component at the particular milestone. Similar to the weightings applied to the milestone components, the summation of the weights applied to the criteria must also equal 1.0. The readiness of each component is the summation of each criterion readiness. Criterion readiness is calculated by the percentage completion of a criterion times the weighting applied to the criterion.

Overall readiness of the milestone is determined by the summation of readiness of each component (derived from its criteria) times the weighting applied to that component for the milestone. Figure 4.1 demonstrates how

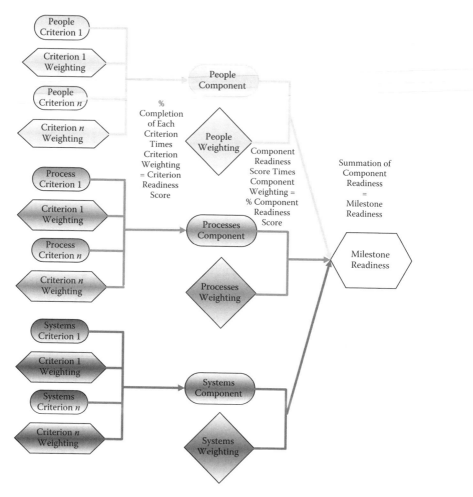

FIGURE 4.1
The readiness process for each milestone.

the readiness of a milestone is derived. Starting initially with the criteria for each component, the percentage completion of each criterion is multiplied by the respective criterion weighting to produce a weighted average readiness score. Summation of the criteria readiness scores multiplied by the component weighting leads to the component readiness. Summation of component readiness leads to the milestone readiness. For each criterion, an objective, quantitative means for assessing percentage completion is required.

Important to the whole process is the weighting applied to the components and each criterion. Weightings are predetermined and are based on the relative importance of the component for the specific milestone. Similarly, criterion weighting is determined by its relative importance to the component at the specific milestone. Often, weightings applied to each criterion are determined by the amount of resources (human effort, time, and money) required to drive the criterion to completion within the allowable time frame as determined by the milestone. We shall see also that as project execution proceeds, the relative weightings of the various components change as the importance of the component changes within the project framework at different milestones (see Table 4.1 for worked example).

Figure 4.2(a) shows the relationship between weighting and the relative importance of each component and criterion-based resource requirements. While a precise relationship may not be required, the graphical presentation is provided to guide project leaders in allocating weights to components and criteria based on relative importance to the particular milestone. Weights are determined and allocated by project leaders with assistance from representatives of the construction, commissioning, and operations organizations who will ultimately be responsible for completing the work associated with each criterion associated with each component. Champions may be assigned for leading component readiness. Component champions are expected to provide objective assessments of the resources (human effort, financial, and time) required in determining the relative importance and weighting for the criterion. Component champions will also be required to steward each criterion to completion within the time frame required for the milestone.

Figure 4.2(b) demonstrates the changing weights allocated to PP&S components as project execution continues over time. As shown graphically, low weights are allocated to people and processes during early project execution since the project is owned primarily by the construction organization, which is busy constructing facility systems. High weights are allocated to system readiness. Over time, as systems are constructed and completed, higher importance and weightings are placed on people and process readiness since these components will be required to ensure a safe and effective handover from the construction organization and the transition through commissioning into continuous operations.

Management and control of the execution stage of the project require project leaders to identify and define several distinct and identifiable targets or milestones. Clearly defined milestones allow project leaders to maintain both the 30,000-foot view and the 200-foot views of the project, which collectively allow for greater control of the execution of the project.

Milestone Determination

Milestones are specific significant events in the project execution stage that indicate that progress has been made toward the final completion of the project. According to Neil Camarta, executive vice president of Suncor Energy

TABLE 4.1

Worked Example of Readiness for Milestone 1 (M1, Ready to Turn Over First System: Industrial and Instrument Air System)

Component Criteria	Criterion Weight %	Percentage Completion %	Criterion Readiness Score %	Component Weight	Component Readiness Score %	Milestone Readiness %
People						
40% of operations workforce hired and oriented	40	80	32			
Standard operating procedures for air system completed	30	85	26	0.2	18	
Six operators trained and qualified to operate air system						
Total	**100**		**88**			
Process						
Critical spares for air system identified and warehoused	50	100	50			
Preventive maintenance schedule defined	20	100	20	0.3	28	87
Computerized maintenance management system enabled	10	70	7			
Contractual service agreement in place and approved	20	80	16			
Total	**100**		**93**			
Systems						
All piping and vessels in place	50	100	50			
System hydro tested	20	100	20			
Commissioning procedures completed	20	60	12	0.5	42	
Compressors commissioned and started						
Total	**100**		**84**	**1**		

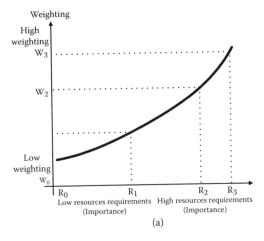

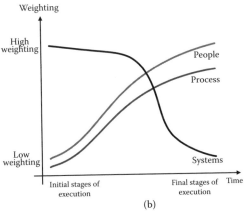

FIGURE 4.2
(a) Weighting and resources requirements. (b) Weighting changes with project execution over time (not to scale).

Incorporated, defining and stewarding to milestones allows project leaders to "eat the elephant one bite at a time" (N. Camarta, personal communication, 2008). For large projects, several milestones are recommended before the project execution is completed. Milestones are determined prior to the start of the execution stage and are generated during one or more working sessions with the project leadership team. Significant achievements or events in the work performed by the construction, commissioning, and operation organizations are identified, and target dates are determined for each milestone.

Some prework is required by each group so that the working session does not become bogged down with working details of each milestone. The project leader or an independent facilitator may be required to assist in identifying and defining *real milestones* and in establishing realistic completion

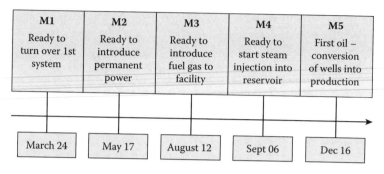

FIGURE 4.3
Typical milestone events for a SAGD operation.

time since there may be a tendency to overidentify milestones and build in contingency for timing of completion. The facilitator is required to be objective and must be able to both assimilate technical information and translate uncertainty among participants into best estimate decisions based on available information.

The deliverable at the end of the working session is expected to be four or five clearly identified milestone achievements or targets with realistic time frames for completion of each milestone. It is important to point out that these milestones must be sequentially identified and must also be sufficiently important to provide a "go-no-go" basis for the decision to proceed from one milestone to another based on the status of completion of the milestone.

The timeline in Figure 4.3 shows a typical milestone arrangement during the execution of a typical steam-assisted gravity drainage (SAGD) project.

Criteria for Readiness

For each milestone, there are certain criteria on which the readiness of PP&S is dependent. These criteria will be defined by the project leadership team in consultation with representatives from the construction, commissioning, and operations organizations. Completing work associated with each criterion allows for progress beyond a milestone to the subsequent milestone. The goals of the various organizations are to identify these criteria and work toward ensuring full readiness by the designated milestone date. Criteria identification and completion are particularly important if a criterion has an impact on the movement of the project beyond the milestone.

Criteria identification requires a working group brainstorming session in which the working group is required to determine all criteria for readiness associated with PP&S for each milestone identified. This process is particularly important for both the commissioning and the operations organizations if there may be a considerable volume of work required to be ready at the milestone date determined. Construction, on the other hand, will usually begin work with a skilled labor force and often with well-developed processes for procurement and work management. Consequently, beyond milestone identification, their criteria for readiness for the above PP&S can be minor.

Criteria for readiness for both the construction and operations organizations require intense working. It is important to point out that some criteria are rooted in legislation and must be completed for progress beyond the milestone. Some of these regulatory criteria are what can be termed *showstoppers*. Showstoppers have the potential to bring the project to a grinding halt in the absence of progress. They must be recognized early and provided sufficient resources and attention to ensure they are addressed adequately in a timely fashion to avoid progress delays from one milestone to another.

In some instances, full criteria readiness may not halt movement beyond the milestone. In such instances, work can proceed with appropriate mitigating actions that will allow completion of the criteria within an acceptable period and with the application of additional remedial resources. Some examples of criteria for readiness are as follows:

People readiness criteria:
1. All personnel are hired and properly oriented.
2. Standard operating procedures (SOPs) are developed for all operating areas.
3. All personnel are properly trained, assessed, and qualified.

Process readiness criteria:
1. The equipment hierarchy has been fully developed for a computerized maintenance management system (CMMS).
2. Preventive maintenance schedules have been fully developed for all critical equipment.
3. Accounting systems have been developed, and all methods for procurement of support and services have been organized.

Systems readiness criteria:
1. The power distribution system has been completed, tested, and energized.
2. The instrument air system has been tested and punch listed, and deficiencies have been identified and resolved.
3. Turnover documentation has been completed inclusive of all quality assurance/quality control (QA/QC) reports.

The Weighting Process

Proper weightings must be allocated to each of the PP&S components at each milestone. More important, the criteria that make up each PP&S component must also be assigned appropriate weights based on relative importance to that component. The weighting process requires appropriate representation from each of the construction, commissioning, and operations organizations, working independently at first to identify and allocate weights to all criteria of PP&S within each organization based on the efforts required to achieve readiness at each milestone. Sharing of this information is required at a project leadership team meeting to collate all PP&S readiness criteria and weighting requirements for each milestone.

Figure 4.4 shows, as project execution progresses across milestone events from M1 to M5, the relative weights for PP&S changes. Although not quantified, the relationship clearly shows that a shift in weighting applies to each component of PP&S as project execution progresses through successive milestone events. At M1, we see high weights applied to system and low weights applied to both people and process. This skewed weighting reflects the reality that maximum effort is applied to construction of the various operating systems that collectively will make up the plant or facility.

By M3, we see a significant shift in weighting applied to all components as people and system readiness become more important in the overall project execution. At this stage, there is significant advancement in construction of physical assets and infrastructure. The need for supporting operating processes begins to become evident (e.g., procurement and invoicing processes). At this stage, also, there is an increasing need for people readiness since it is possible that completed systems can be turned over to the operations group, and continuous operation may be required for ongoing construction and commissioning activities. For example, early system turnover to operations may include the instrument/industrial air system or the power distribution system. Turnover of both these systems has cost-saving potential as the reliance on portable generators and compressors for supporting construction and commissioning activities can now be replaced by facility or project supplies.

At M5, the weighting has shifted significantly to reflect the high importance placed on both people and process readiness relative to low weighting on system readiness. At this stage, execution has advanced to the point at which the facility can be started up and put into continuous operation. People readiness and the availability of supporting processes such as accounting and procurement now take prominence and are therefore highly weighted. At this stage, the expectation is that all systems are essentially fully completed with minor deficiencies. These deficiencies are clearly listed and will not have an impact on personnel safety or production. Here, also, it is crucially important that personnel are properly trained and ready to assume

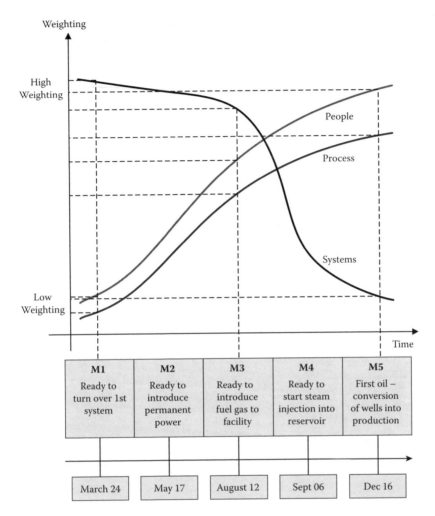

FIGURE 4.4
Weighting changes with milestone events.

responsibilities for continuous operations. Moreover, appropriate processes for continuous operations have been developed and communicated.

Who Is Who in the Readiness Process?

The three main organizations involved in the readiness process are the construction, commissioning, and operations teams. Each team will have personnel responsible for specific criteria deliverables at each milestone.

Assignments and deliverables must be clearly communicated and supported for each milestone event. Each team member with assigned responsibilities must understand his or her role and how his or her deliverables fit into the milestone deliverables. More important, the team member must be empowered to take necessary steps in delivering results and must also be held accountable for poor performance. It is important to point out that the component leaders play a crucial role in guiding the PP&S team toward milestone deliverables.

Construction Organization

All members of the construction organization work together to meet the system requirements for each milestone identified. Leaders of all disciplines in the construction organization shown in Figure 4.5 must work together

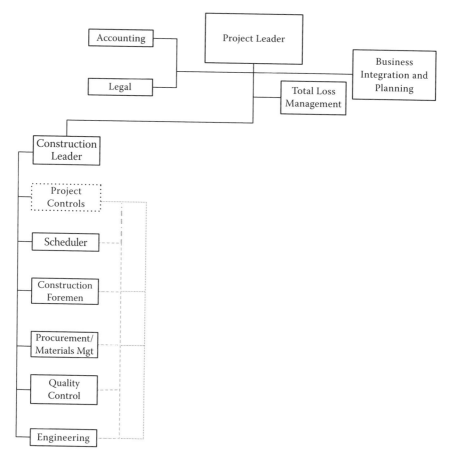

FIGURE 4.5
Construction organization and discipline leads.

to deliver milestone deliverables at the milestone time. Early identification of issues, challenges, or concerns that may have an adverse impact on the milestone is required from each discipline lead and must be communicated to the construction leader once identified. Weekly progress updates on the readiness of the PP&S for the milestone will identify gaps and weaknesses in related criteria. With early identification, mitigation actions can be initiated to maintain the milestone deliverables according to the schedule.

Commissioning Organization

Like the construction leader, the commissioning leader is responsible for the efforts and deliverables of the commissioning organization for each milestone event. All disciplines within the commissioning team are responsible for their milestone criteria deliverables according to the schedule. At each milestone, the commissioning organization works toward ensuring proper system turnover. All systems must be tested and verified ready for continuous operations by the operations organization. Figure 4.6

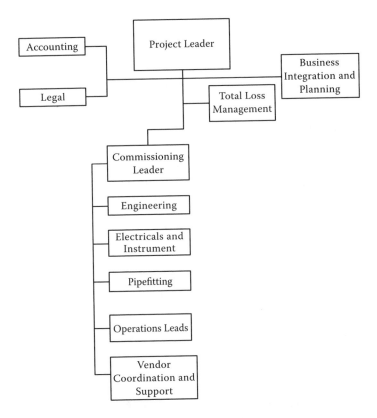

FIGURE 4.6
Commissioning organization and discipline leads.

provides an example of a typical commissioning organization on an industrial project.

Issues or concerns that may hinder or impede milestone achievement must be brought to the attention of the commissioning leader so that they can be elevated for mitigation and corrective actions as quickly as possible.

Operations Organization

The operations leader assumes a huge responsibility during project execution. While it is the last organization of the project execution team to be formed, the responsibilities within the operations organization grow with progressive milestones. This organization prepares for steady-state operations and has tremendous potential to delay the project if people and processes are not ready when the construction and commissioning organizations are ready to turn over completed and tested systems to the operations organizations. Figure 4.7 provides an example of a typical operations organization on an industrial project. Early identification of issues and concerns that may prevent people and process readiness is necessary. Once informed, the operations leader must preferentially allocate resources to areas of weaknesses and gaps to ensure that the project execution remains on schedule.

Table 4.1 provides a worked example of readiness for milestone 1 (M1), ready to turn over the first system. In this example, I assume turnover of the instrument and industrial air system for an industrial facility. Criteria for PP&S readiness were defined and the overall readiness of the milestone event computed based on the assumed weights and percentage completion of each criterion.

In this example, a weight of 0.2 was placed on the people component; weights of 0.3 and 0.5 were assigned process and system components, respectively. Using the readiness method shown in Figure 4.1, $(C1 \times W1) + \ldots + (Cn \times Wn) =$ Component readiness score. The people readiness score is therefore computed as follows: $(0.4 \times 80) + (0.3 \times 85) + (0.3 \times 100) = (32 + 26 + 30) = 88\%$. Similar computations yield process and system readiness scores of 93% and 84%, respectively. Milestone readiness is calculated as follows: (People component readiness score × People weight) + (Process component readiness score × Process weight) + (System component readiness score × System weight). Substituting values accordingly:

M1 Readiness = $(88 \times 0.2) + (93 \times 0.3) + (84 \times 0.5) = (18 + 28 + 42) = 87\%$

With this state of readiness in mind, the project leadership team can make an informed decision about whether to advance work toward the next milestone, declaring M1 completed. Residual resources will therefore be made available for full completion of M1 PP&S criteria. Occasionally, a readiness state closer to 100% is required to enable movement to the subsequent

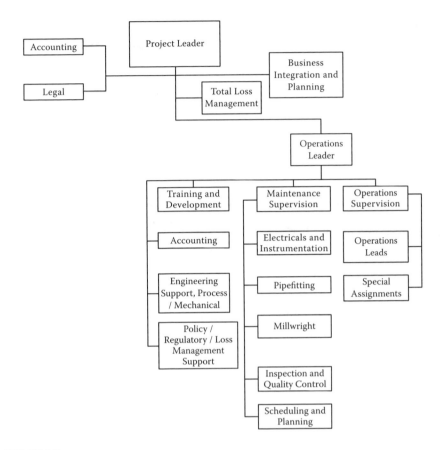

FIGURE 4.7
Operations organization and discipline leads.

milestone; however, these decisions will be at the discretion of the project leadership team.

Pulling It Together

Critical to success in project execution is that all information must come together in a controlled fashion and at a defined and regular frequency. Progress updates on a continuous basis are required to ensure that project execution continues consistent with milestone deliverables for meeting the schedule and budget. The project leader or the leader's delegate is required to provide status updates on a regular basis to corporate leadership. Initially, biweekly status updates may be adequate to ensure that the project execution continues to progress according to plan. As the project approaches peak activities, the frequency of updates must be increased to weekly.

Weekly updates appear to be optimal since they allow workers in each organization to get back to the workplace and actually achieve progress toward deliverables between updates. More frequent progress updates become counterproductive as discipline leads spend too much time on preparing for update sessions as opposed to working toward deliverables for each milestone event. Occasionally, an update on a significant criterion may be required before the scheduled update. This can be achieved via a special meeting between the assigned discipline leader or organization leader and the project leader.

Prior to the scheduled meeting, the construction, commissioning, and operations leaders are expected to provide numerical estimates of readiness on criteria deliverables for each component of PP&S components for the specific milestone so that overall milestone readiness can be computed. This approach to information updating ensures that the meeting is an information session rather than a working session. It is also important that the meeting is properly managed relative to a fixed agenda. A strong chairperson is required to keep the meeting on track, assign work, take corrective actions, and make critical decisions regarding course corrections when necessary. The chairperson may often be the project leader or a delegate.

Empowerment and Accountability

The readiness process requires that leaders be empowered to achieve milestone deliverables. Responsible personnel within each organization must be given the authority and resources to accomplish the tasks assigned for each milestone and held accountable for the performance of their work groups. During project execution, there must be low tolerance for insufficient effort since poor performance by one organization or work unit can jeopardize the efforts of the entire project execution team. Each member of the respective organizations must clearly understand the importance of work performed by his or her organization and how his or her deliverables have an impact on the overall schedule or budget during execution.

The project execution stage is highly demanding, and innovative methods for delivering results on time and within budget are required. Employees must be allowed to make decisions within reason without being tied to bureaucracy in the decision-making process. An effective management of change (MOC) process is required since deviations from originally planned activities on each criterion may result in unintended consequences. An MOC process assists decision makers in identifying all issues and concerns that may arise from making a change or deviating from planned activities. The project leadership approval may be required for changes since deviations from plans may introduce externalities not planned for by other organizations. New risks and hazards may be introduced into the process by making a change, and these undesirable externalities must be stewarded and mitigated to manageable levels to avoid undesirable outcomes.

Prior work experience in project execution and the ability to work flexible and long hours become important assets provided by the employee. It is important that high stress does not get in the way of productivity, and corporate assistance in dealing with work stress must be provided. With extended work days, weeks, and months, careful attention to stress management and burnout is required to maintain high productivity and quality control on work.

The Go–No-Go Decision

Progress from one milestone to another depends on the state of readiness of each milestone. In the real world, it is entirely plausible that a milestone will not be 100% ready, and the decision to proceed to the next milestone will be necessary. Here, the concept of the *go-no-go* decision will become important. The project leader is required to make objective decisions to proceed to the subsequent milestone without full completion of the prior milestone.

Generally, milestone criteria that trigger this decision are those associated with people readiness (e.g., status of training or availability of SOPs), regulatory requirements, and system completion based on construction issues. In most instances, however, the project leader may decide to progress to the next milestone while enhanced remedial efforts and resources are applied to bring the criteria to readiness in the shortest amount of time possible.

5

People Readiness

Introduction

People readiness during project execution is primarily about having qualified and competent workers to support the movement of the project from one milestone to another and for supporting the project during continuous operations. People readiness during project execution is a critical requirement for projects to be completed within budget and on schedule. This is because people readiness has the potential to derail the project at any milestone based on the state of readiness of the people component. Incompetent or unqualified personnel due to inadequate training present both legal and moral dilemmas.

If people are not trained and qualified adequately at each milestone, the risk exposures of the organization are increased beyond the acceptable tolerance of the organization. As indicated in Chapter 4, people readiness can influence the go-no-go decision and often requires special influence with mitigating actions before a project leader may proceed beyond the milestone. People readiness becomes increasingly important during later milestones, for which higher weights are assigned to people readiness. People readiness is primarily about the operations organization since both the construction and commissioning organizations tend to be transient, generally moving from project to project.

The construction organization is often contractual and operates as a completely independent organization, looking after its own training, safety, and wage issues. In addition, the construction organization is generally about getting in, getting the job done, and moving on to the next job. Often, the construction organization is brought in when needed as a complete organization equipped with all infrastructures required to support itself in ensuring its people readiness. In remote areas, this also includes camp accommodation and an appropriate work schedule to best meet the needs of the project. People readiness for the construction organization is often a nonissue. Although this is the case, it is important that contractual obligations of the construction organization with regard to people readiness are clearly defined up front so if costly issues do arise, there are no questions regarding who the responsible party is.

People readiness in the commissioning organization is often a nonissue also. There is often an abundance of commissioning expertise readily available in the external market. Moreover, there appears to be a strong network of commissioning expertise since commissioning personnel travel from project to project, performing their roles before moving to the next work assignment. Occasionally, the commissioning organization is structured in a similar manner to the construction organization and can be parachuted in as required.

The operations organization is generally the one in which people readiness is of greatest importance. This is because of the need to train and retain personnel to assume the operating responsibilities for the project on a continuous basis once turned over to operations. People readiness requires significant levels of preparation in recruitment and selection, relocation (if necessary), training and competency, and retention.

People readiness in the operations organization becomes more important when technology is new and on greenfield projects with new designs. In addition, if the location of the workplace is remote, there may be greater difficulties in securing and retaining the services of skilled and experienced personnel. People readiness begins at the recruitment and selections stage, or pre-M1 (milestone 1), and progresses through to a trained, qualified, and competent operations staff by the time the project is ready for steady-state operations at M5.

Building the Operations Organization

The operations organization starts with recruitment and selection of suitable candidates for key leadership roles in the operations organization. These key leaders will have responsibilities for identifying people readiness criteria within each discipline and for ensuring people are hired, relocated, trained, and qualified (people readiness) for each milestone activity.

Figure 5.1 shows the sequence for filling key leaders in the operations organization. Recruitment and selection of key personnel shown allow the operations organization to develop to meet the milestone requirements for people readiness as the project proceeds.

Among the first roles filled should include the operations and maintenance supervisors who will assist in the selection of skilled and experienced personnel for building the operations organization. Recruitment for training and development, accounting, policy, regulatory, and loss management support roles is done at the same time to allow for the formation of the core leadership of the operations organization. The training and development role is of particular importance since this individual is responsible for ensuring that personnel are trained consistent with milestone requirements. This group forms the core leadership team. This core team is also responsible

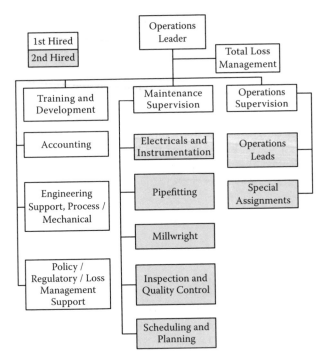

FIGURE 5.1
Hiring sequence of key personnel in the operations organization.

for developing criteria for people readiness and the required processes for ensuring that the project can be operated on a controlled, safe, and sustained basis. Essentially, all members of the core team have key criteria responsibilities in meeting milestone readiness at each milestone. They are instrumental in defining the framework for seeing each criterion to completion.

The second group to be hired forms the second-level leadership and is essential in executing plans for ensuring readiness at each milestone. These personnel will assist in fine-tuning the plans required to meet criteria completion for each milestone for people, process, and system readiness. Finally, frontline personnel are hired (not shown in Figure 5.1). These operations personnel are required to keep the plant operating on a continuous and sustained basis after they are properly trained and all criteria for people readiness have been completed to 100% for full people component readiness.

Criteria for Readiness

At each milestone, there is a different list of criteria for people readiness. In addition, there may also be people readiness criteria that extend across several milestones. I identify some criteria for each milestone for a typical steam-assisted gravity drainage (SAGD) operation. Table 5.1 shows people readiness criteria across M1–M5 associated with an SAGD operation. Criteria identified for all components in the readiness process should be measured and quantified in some objective and quantitative manner. Criteria identified in Table 5.1 for people readiness seek to reflect specific elements of measurability so that readiness can be computed consistent with the readiness process. The list may not be all encompassing but is sufficient to capture and address those variables likely to affect the readiness of personnel as the project proceeds. Weights are applied to each criterion based on the relative importance of the criterion as determined by the working organizational leadership team.

Training and Qualification

Training and qualification of all personnel are among the most important criteria across all milestones during the execution stage of the project. Training and qualification go beyond the need to understand the operation of the plant or facility. Personnel must also be trained on how to respond in the event of emergencies. They must be trained to work under adverse conditions in the event of gas or vapor release and how work is undertaken in confined spaces.

Training and qualification can occur internal to the organization if the training capability is available or by utilizing external resources. Skilled and competent training experts should always be used for conducting training for best performance. Poor-quality and poorly delivered training leads to weak understanding and learning by trainees, and it will be necessary eventually to repeat the training exercise. For regulatory training, it may be logistically more manageable to have personnel trained by expert institutions to ensure full understanding and competency. In addition, expert organizations lend credibility to the training process in the event of a mishap. Qualification requires an effective mechanism designed to ascertain that the trainee has a good grasp of the information presented. The qualification process can take the form of a simplistic classroom testing process to the advanced combinations of both written testing and a field demonstration of the practical application of the information presented.

Some regulatory certification may have fixed expiry dates and duration, and one should seek recertification before expiry to ensure personnel are compliant. More important, records of training must be maintained for all personnel so that people readiness can be verified in an auditable framework. Well-maintained training records are required to ensure that only qualified personnel are allowed to perform work requiring certification.

TABLE 5.1

People Readiness Criteria across Milestones 1–5 for an SAGD Operation

Pre-M1, Move to Project or Plant Site	M1, Ready to Turn Over First System, March	M2, Ready to Introduce Permanent Power to All Systems, May	M3, Ready to Introduce Fuel Gas to Facility, August	M4, Ready to Start Steam Injection into Reservoir, September	M5, First Oil: Conversion of Wells into Production, December
Adequate accommodation and office space available for core personnel in the operations organization. Core operations leadership team hired and relocated to project or plant site areas, inclusive of families where necessary. Assumes remote locations. Adequate transportation to work site available if remote site.	Operations leads hired and oriented. 40% of facility operators hired and oriented. List of standard operating procedures (SOPs) defined for facility. Operating manual completed and available for personnel training. Critical practices developed for this system. SOPs completed for first system. Required operators hired, trained, and qualified to operate the first system. Maintenance lead personnel hired and oriented. (Assumes certification is adequate for performing work on the system equipment with minimal training.)	Codes of practices for zero-energy lockout completed. Operators trained and qualified on zero-energy lockout. Electrical and instrumentation personnel trained and qualified on the power distribution system. SOPs for power introduction and power distribution prepared. All other SOPs at least 40% completed. Operating areas defined and equipment list prepared for each area. Operations manuals for each operating area completed. Operators trained and competency assessed in use of SOPs for power introduction to facility and power distribution.	All operators required for continuous operations of the facility hired and oriented. All lead operators and at least six operators trained in operating the following systems: • First system turned over • Power introduction and its distribution • Fuel gas and flare systems Codes of practice governing hydrogen sulfite (H_2S) and confined space entry completed. All personnel (construction, commissioning, and operations) trained in hydrogen sulfite awareness.	All SOPs completed. Plant risk ranked and operating personnel assigned based on skills. Skilled and experienced operators assigned to areas of high risk. Shift system developed and schedule determined. Operators assigned to shift based on skills, experience, and overall strength of shifts. A balance of strength across all shifts is required. All operators have received operating manual for their assigned area. They have also received training in their assigned area.	All personnel are qualified to operate assigned areas. Competency assessed. All maintenance personnel are hired and oriented. Operations organization is complete.

Training records prevent costly duplication of training and provide due diligence support for the organization in the event of unfortunate and unforeseen incidents. In addition, while uncommon, some workers will consider shirking on the job under the guise of not being trained and qualified to perform a particular task. Good records can either support or refute the claim. On a new project site with new technology, the learning curve for most personnel will be steep, and a lot of hard work is required. A comfortable learning environment will be required, and there may be the need to conduct repeat training sessions.

When the technology may already be in existence, it may be possible to have personnel trained at other operating sites before transfer to the project site. Training at another site does not void the milestone requirements for trained and qualified personnel. However, hands-on training at another operating facility may enhance the capabilities of a trained and qualified employee such that when the employee is transferred to the project site the learning curve will be of a much smaller gradient.

Figure 5.2 provides an overview of some of the training required for people readiness in a typical industrial facility. Although the list is not complete, it identifies most areas of training required and categorizes training under the broad headings for which new employee training will be required. Figure 5.2 can be expanded for identifying training required for people readiness consistent with the type, location, and scale of any industrial project.

Standard Operating Procedures (SOPs)

Standard operating procedures are designed to guide operating personnel in performing work safely. SOPs provide a sequential set of steps to operating personnel such that when followed properly, a worker who is not familiar with the equipment or process will be able to safely and effectively perform an assigned task. If new equipment and machinery are involved in the process, vendors may provide the accompanying operating manuals.

The operations organization may be required to take the operating manuals and condense them into SOPs that can be used by operating personnel. Appendix 1 provides an example of a well-written standard operating procedure developed for regenerating a weak ion, cation exchanger involved in the water-softening process for an industrial boiler or steam generator. A SOP will typically provide the user the following details:

1. A reference number and title for identifying and searching for the procedure
2. Date a procedure was written

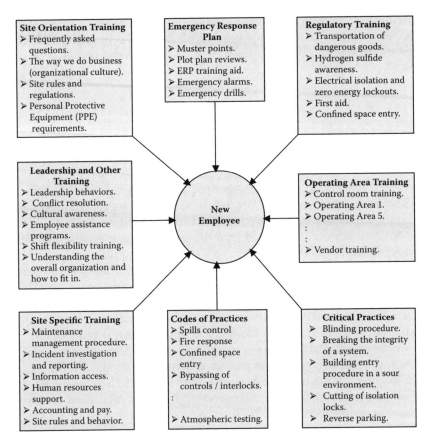

FIGURE 5.2
Typical training requirements for people readiness in an industrial facility.

3. Author of the procedure

4. Person(s) who reviewed the procedure

5. Revision number of the procedure

6. Page numbers of the procedure (normally numbered in the format page *x* of *y*)

7. The steps involved in starting up, shutting down, or operating the equipment or machinery

8. Identification of hazards

9. Precautionary measures and personal protective equipment (PPE) required.

Occasionally, an SOP may also include a hazard analysis assessment and recommendations for addressing each hazard identified. All steps in an SOP must be followed as outlined in the procedure. When difficulties in following the procedure are encountered or a step is no longer valid, an appropriate level of authority is required to make the required changes to the procedure, with follow-up training for all personnel involved in using the SOP.

Critical Practices

Critical practices are types of SOPs that require personnel to follow precisely the directions provided in the procedure. Critical practices are designed to ensure the safety of personnel, equipment, and machinery. Failure to follow the steps and controls identified in a critical practice may result in severe injuries or fatality to personnel or severe damage to equipment and machinery. In all work environments, safety standards that emphasize that personnel safety comes above all else are absolute requirements.

Inexperienced workers should not be allowed to undertake a task defined as a critical practice without adequate supervision. *Full competency is required* for all employees performing work that requires critical practices for performing tasks associated with the work. Appendix 2 provides an example of a critical practice. The example provided demonstrates the steps necessary for unloading a truckload (tanker) of acid into site storage tanks of a facility. Full and uncompromising adherence to the steps identified in a critical practice procedure is required by the organization and personnel using the critical practice.

Codes of Practices

Codes of practices (COPs) are generally derived from regulatory requirements that guide operations within a region, province, state, or country. COPs are intended to establish the minimum standards of safety or environmental protection required by regulatory bodies and institutions for the region. Most organizations will seek to establish standards that exceed the standards defined by regulators.

Some common COPs in an industrial operation may include those governing entry to confined spaces, energy lockout, fall protection, and emissions control. Appendix 3 provides an example of an energy lockout COP. Full and uncompromising adherence to the steps identified in a COP is required by the organization.

Training, Qualification, and Competency Matrices

Beyond the need for well-maintained training and qualification records, there is the need for visible and updated training information that can be reviewed quickly to assess progress with training, qualification, and competency for all personnel and people readiness. The use of a training matrix is an easy method for conveying this information. A training matrix provides a snapshot of all groups at the project site, the training required, and the status of each individual's training at a point in time. Clearly identified also are personnel who are designated capable of qualifying trained personnel and assessing the competency of those trained.

Figure 5.3 shows a typical training, qualification, and competency matrix. Using a color scheme, it is possible for project leaders to assess at a glance who are trained from those who are not. More important, project leaders can easily quantify the status of training as a criterion for people readiness on an ongoing basis. Using a training matrix, training can be easily expressed as a percentage completion for assessing overall readiness at each milestone.

The matrix identifies required training for each person. It also identifies trainers and can show the date training was undertaken. Figure 5.3 identifies areas in which workers have been trained and qualified by competency assessors. Discipline experts are identified and are required to conduct training and assess the competency of personnel before they are deemed qualified. Discipline experts may include mature and experienced engineering experts, operations lead personnel, advisors, commissioning experts, and external vendors. These personnel may be required to qualify both operations and maintenance personnel.

Often on projects built on new technology, personnel may be qualified to operate the system but may require operating experience to be deemed competent. In such situations, it may be necessary to maintain the training and competency matrix beyond turnover of the project to the operations organization and after the project moves to steady-state continuous operation.

A visible training matrix with vibrant color schemes to represent untrained workers can advise project leaders at a glance of looming people readiness issues. Visible colors easily attract the attention of leaders so that timely intervention can be made to avoid derailment of the execution process. Resource reallocation or increased resources can be provided to avoid any delays at milestones. Ownership of the training and competency matrix resides within the operations organization but must also be visible and available to all organization leaders. Changes to the matrix should be made by one individual only, usually the training and development leader. A single point of control and ownership is required to avoid confusion and delays in updating the matrices.

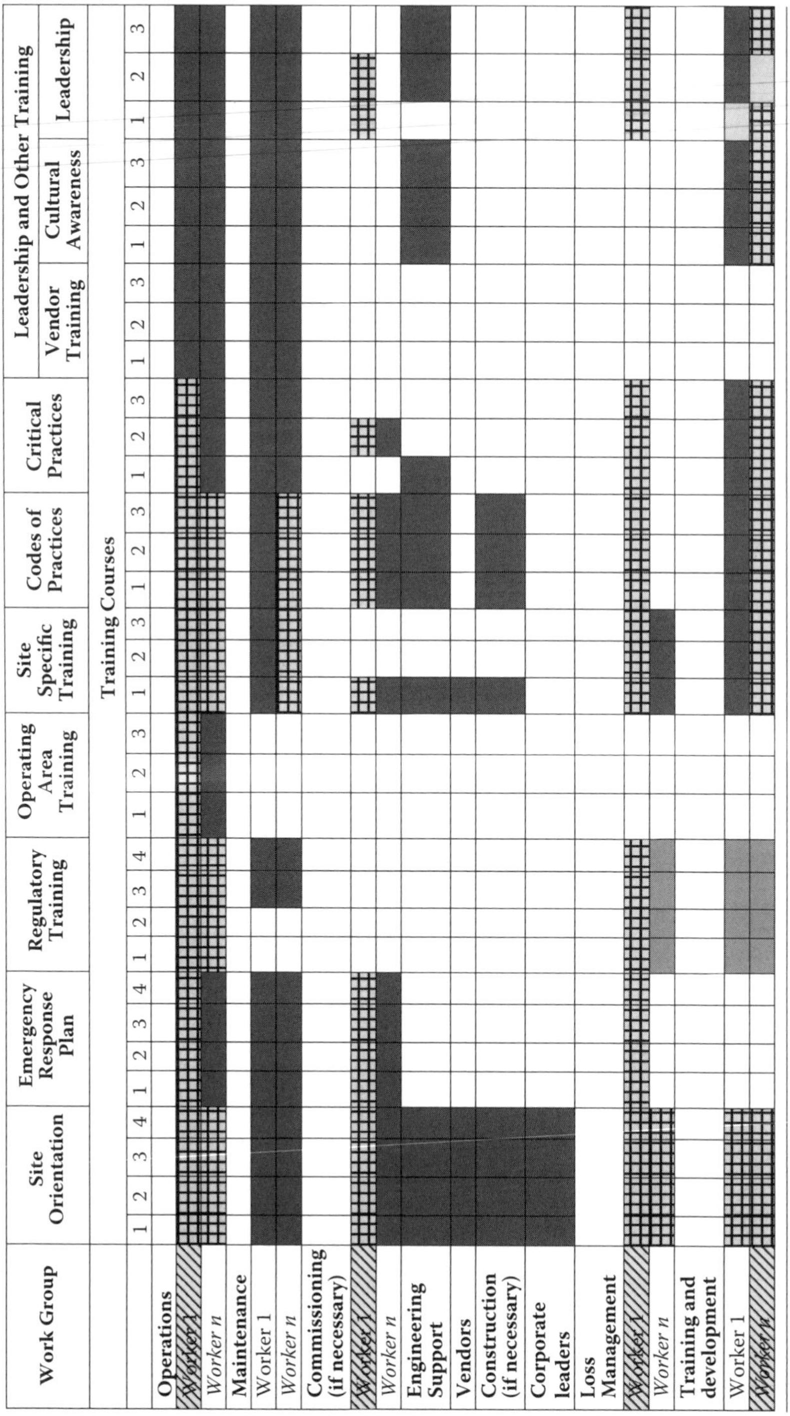

Work Group	Site Orientation				Emergency Response Plan				Regulatory Training				Operating Area Training			Site Specific Training			Codes of Practices			Critical Practices			Vendor Training			Cultural Awareness			Leadership		
																Training Courses												Leadership and Other Training					
	1	2	3	4	1	2	3	4	1	2	3	4	1	2	3	1	2	3	1	2	3	1	2	3	1	2	3	1	2	3	1	2	3
Operations																																	
Worker A																																	
Worker n																																	
Maintenance																																	
Worker 1																																	
Worker n																																	
Commissioning (if necessary)																																	
Worker A																																	
Worker n																																	
Engineering Support																																	
Vendors																																	
Construction (if necessary)																																	
Corporate leaders																																	
Loss Management																																	
Worker A																																	
Worker n																																	
Training and development																																	
Worker 1																																	
Worker n																																	

FIGURE 5.3
Training and competency matrix.

Training and competency records must also be provided when changes to the training matrix are required. All training records must be owned and maintained by the training department. It is important, therefore, that training and qualification records are provided to the training department on an ongoing basis along with requests to update the matrix. Training and qualification records must include details regarding the course content, the employee signature, the trainer's signature, and date and time of training. For group training, participation records and group signatures will be required for auditing purposes. Figure 5.4 provides a simplified training record of competency assessment to be retained on file.

Employee Name:		*Operating Area / System*	
Job Title:			
Assessor Name:		Assessment Date:	

Employee Assessment

1) System or Area Description:			
2) Evidence Presented (attach copy where applicable):			
Performed by Supervisor	Comment	Feedback and Other Supporting Information	
➢ Observation of employee doing task as per Critical Practice & SOPs. ➢ Assessment of outcome of employees work. ➢ Task discussions and review of questions. ➢ Discussions on responses to simulated conditions. ➢ Work related assignments.		➢ Feedback from peers and other workers. ➢ Team outcomes where the individual contribution is evident. ➢ Record of work or training activities.	
3) Employee Comments:			
Candidate's Signature:			Date:
Assessor's Signature:			Date:
To be forwarded to Training Department upon completion.			

FIGURE 5.4
Competency assessment form. (© Suncor Energy Inc. With permission.)

Push/Pull Effect of the Training Qualification and Competency Matrix

During project execution, training of workers is a critical component of people readiness. In situations of budgetary constraints, training tends to assume the role of a nice-to-have as opposed to a need-to-have requirement. This, however, is an *erroneous stance* to take with regard to training. Effective project leaders will recognize that trained and competent personnel not only increase productivity but also reduce risks, incidents, and rework. Indeed, skilled and competent personnel improve the overall situation in case of budgetary constraints by reducing the likelihood of incidents, injuries, poor performance, rework, and damage. Project leaders must move beyond the view that training will be accomplished when personnel are free from other work. Quite possibly, during the project execution stage, training and development generate one of the highest returns on investment.

Within work groups and divisions, an updated and visible training and competency matrix forces department leaders to actively *push* personnel into training when the contrast between *training required* and *competency confirmed/qualified* on the matrix shows the wrong balance. Workers may also approach department leaders to be sent for training once they recognize a training gap associated with their names. Furthermore, a contrast between training and qualification among various work groups encourages leaders to more aggressively *push* personnel into training.

An updated training and competency matrix strategically located where workers gather (e.g., lunchrooms or conference rooms) has the effect of encouraging untrained and nonqualified personnel to actively seek training. This is considered the *pull effect*. There is a genuine desire by all employees to seek training to improve their capability at the facility. In some instances, peer pressure may also have the desirable effect of encouraging delinquent personnel to actively pursue training and competency in areas that are required. This is particularly so in high-risk work for which the health and safety of all workers is dependent on the competency of all workers.

Competency Assessors

Competency assessors are required to conduct training and assess the capabilities of personnel (qualify) before they are deemed qualified to work. Competency assessment is a necessary part of the people readiness process. Competency assessors themselves must be competent in the subject they are required to qualify and assess personnel on their abilities to perform assigned work. To fulfill this role, competency assessors must possess a high level of expertise in the area of training and must be approved by the operations organization core leadership team to perform this function. Competency assessors are required for two reasons:

- Competent trainers and assessors provide the best opportunity for well-trained and qualified personnel. Well-trained operations personnel provide the best opportunity for reducing the likelihood of incidents, injuries, poor performance, rework, and damage and maintaining and sustaining continuous operation.
- In the event of an incident or accident, during an investigation, if the competency of personnel is questioned, credibility is enhanced when expert trainers have conducted the training and qualification of personnel involved in the incident.

Competency assessors are generally selected from trained, qualified, and competent experts in operations, controls, engineering, commissioning, and vendor support. It is important that the qualification or competency assessment process be objective and complete. Lack of objectivity and completeness in the competency process can result in a poorly skilled workforce with potential adverse consequences for critical work and assignments. Workers may be hurt or injured in the absence of an effective competency process, and equipment, machinery, and the environment can be damaged.

6

Process Readiness

Introduction

Process readiness is an integral component of the overall project execution stage. Process readiness focuses on ensuring processes are in place to allow the continuous operation of a facility on a sustained basis. Generally, process readiness will be the responsibility of the maintenance leaders within the operations organization. The maintenance leader develops processes to ensure all assets are adequately cared for in a proactive manner and can respond to unplanned outage situations. This chapter provides an introduction to process readiness for a typical industrial plant or facility.

Computerized Maintenance Management Systems

A computerized maintenance management system (CMMS) is intended to manage the workflow process of maintenance activities in a facility with a lean workforce. A CMMS is an effective cost-saving approach to equipment and asset maintenance. Having an effective CMMS is an absolute necessity for any industrial facility or operation. A good CMMS program allows the operations organization in a facility to perform proactive inspections and monitoring of equipment and assets to improve overall reliability, availability, and operating efficiency. Some CMMSs operate on the basis of fixing the equipment or machinery when it is broken. Tinham (2008) advised that "break/fix—often thought of as the simplest and cheapest, since there's no inspection workload—is ultimately the most expensive" (p. 20). A good CMMS system is one that is user friendly and allows maintenance personnel to proactively schedule maintenance work on equipment and assets to prolong the operational excellence of the equipment, assets, and facility.

The CMMS must be user friendly to allow operating personnel to easily communicate abnormal operating observations on equipment and assets so that a maintenance workforce can respond to these observations before damage or failures occur. A priority notification also assists maintenance personnel in the response sequence for maintenance on equipment and assets. For example, a leaking pump seal in a water service may be at a lower priority than a strange humming sound coming from a similar pump in a hydrocarbon service. A selected priority with sufficient description from operating personnel should provide maintenance personnel enough information to effectively prioritize work among competing priorities and in situations of limited worker availability. Care must be taken to educate plant or facility operators on operating conditions that genuinely constitute high priorities. Failure to do so can result in situations in which every entry may be classified as a high priority, thus rendering the prioritizing process ineffective.

An effective CMMS provides sufficient information to end users such that users can select equipment based on the operating area within which the equipment resides, the equipment types, the specific piece of equipment, and further drill-down requirements as necessary. A numbered system is often used to ensure adequate information is available for identifying equipment and machinery at the facility. Figure 6.1 shows drill-down possibilities for an effective CMMS.

Criticality of Equipment and Development of Work Packages

A requirement for process readiness is the identification of the criticality of specific pieces of equipment and machinery with the development of appropriate work packages for maintenance of this equipment in the event of unplanned outages. A risk matrix is necessary to establish the risk ranking or criticality of the equipment. The risk ranking of the equipment determines the level of preparedness required for maintaining the equipment. Figure 6.2 provides a simplified version of a risk matrix that can be used in determining the risk ranking of the equipment.

Criticality of equipment is determined by the risk exposure brought about by failure of the equipment. Using the probability of a failure occurring within 1 year and the severity of the impact of the failure for a particular piece of equipment, a criticality ranking due to the risk exposure can be determined for the particular piece of equipment. To mitigate risk exposures and outage time associated with failure, work repair packages and spare parts inventories can be developed.

An example of a critical piece of equipment is a warm lime softener (WLS) for water treatment in a recycle water service with a throughput of

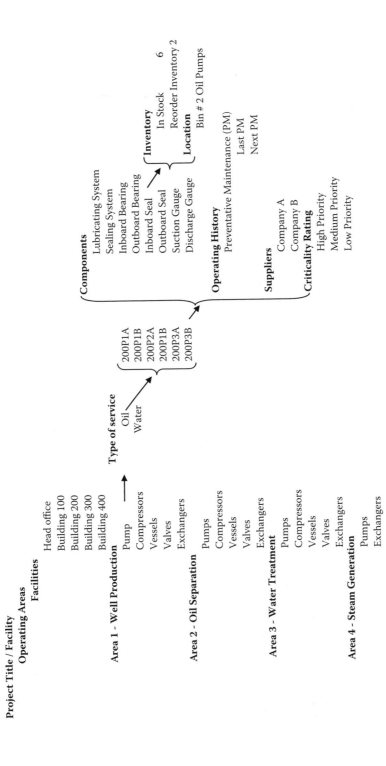

FIGURE 6.1
A simplified computerized maintenance management system (CMMS).

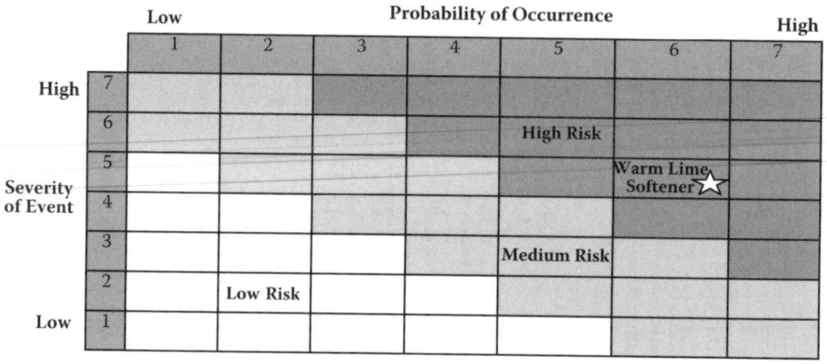

FIGURE 6.2
A simplified risk matrix used in determining the criticality of equipment.

500 M³/hour with no backup or alternative system and boiler feed water storage limited to 4,000 M³. Failure of the system will result in a plant or facility shutdown within 10 hours if the plant consumes 400 M³/hour of water. While such systems are very reliable, the likelihood of failure within 1 year is real, and the impact on operations is total facility failure within 10 hours. Figure 6.2 is the risk ranking of the WLS failure as indicated by the star. Once the criticality of the equipment is determined from its risk ranking (High risk = Critical), maintenance work packages can be prepared for all aspects of repairs and preventive maintenance.

Work packages will include a list of all activities to be performed when undertaking repairs or preventive maintenance. The work package will also identify who (contractor or owner) conducts the work, types of skills and expertise required, maintenance equipment required, and a procedure for performing repairs or preventive maintenance.

For example, repairs of a WLS used in the example provided from an unplanned outage due to a damaged internal rake may have a typical work package as shown in Table 6.1. An internal rake in a WLS is a mechanical system at the inside of the WLS that is used for gently moving the sludge produced in the water-softening process toward the center of the WLS. The internal rake prevents the sludge from becoming dense so that water introduced through a central pipe close to the bottom of the vessel can be filtered through a light sludge bed, thereby promoting chemical softening reactions in the water-softening process. Table 6.1 identifies the work activities to be conducted and the service providers for each work activity for completion of repairs. Some aspects of the work may require the use of detailed procedures and codes of practice, such as a confined space entry procedure or code of practice as determined by the organization.

Determination of equipment criticality and construction of work packages for such critical equipment allow for a smooth and effective maintenance management process that is highly efficient. Such work packages provide

TABLE 6.1

Typical Work Package for Repairs due to Unplanned Outage of a Warm Lime Softener

Work Activities	Trades Required/ (Who)	Contractor/Service Provider	Resources Required	Expected Duration	Backup Service Providers
Remove water from vessel (pump water to storage).	Pipe fillers and mechanical trades Operating personnel/ supervision	In-house operating personnel Maintenance Support	Temporary pump and fittings to support pumping out of vessel	Calculated based on temporary pump capacity and volume of warm lime softener	Contractor A for supplying pumps and mechanical trades
Cool down vessel (remove vents doors and installs cooling fans).	Pipe fillers and mechanical trades Operating personnel/ supervision	In-house operating personnel Maintenance Support	Air movers and cooling fans	15–20 hours estimated to reduce temperature of vessel to permit work	Contractor A for supplying pumps and mechanical trades
Remove sludge and haul to landfill sludge storage. No vessel entry permitted.	Skilled vacuum truck operators and support staff Operating personnel/ supervision	Contractor B: Vacuum supplies and cleaning services; number of trucks and personnel defined to ensure continuous operations	Vacuum trucks High-pressure hoses and flush pumps	Calculated based on expected volumes of sludge	Alternative suppliers: 1–3 identified
Clean warm lime softener and prepare for vessel entry.	Skilled cleaning professionals with confined space entry certification Operating personnel/ supervision	Contractor B: Vacuum supplies and cleaning services; numbers of trucks and personnel defined to ensure continuous operations	Vacuum trucks High-pressure hoses and flush pumps Confined space entry professionals and safety watches	Estimated or based on industry standards	Alternative suppliers: 1–3 identified

Continued

TABLE 6.1 (*Continued*)

Typical Work Package for Repairs due to Unplanned Outage of a Warm Lime Softener

Work Activities	Trades Required/ (Who)	Contractor/Service Provider	Resources Required	Expected Duration	Backup Service Providers
Internal inspections and assessment.	In-house maintenance team Operating personnel/ supervision	Not required	Scaffolding	3–4 hours	Not required
Perform mechanical repairs to damaged rake and internals.	In-house maintenance team Operating personnel/ supervision	In-house maintenance team; continuous operation to minimize outage	Cutting/welding expertise	Based on assessment of work and repairs to be undertaken	Alternative suppliers: 1–3 identified
Quality assurance/ control. Button up system, test, and return to service.	Pipe fillers and mechanical trades Operating personnel/ supervision QA/QC inspection personnel	In-house operating personnel In-house QA/QC personnel Maintenance support	Internal services	Based on start-up procedure and SOP recommendations	Not required

Note: Work activities and work package may change based on seasonal conditions. For example, winter repairs may require scaffolding for hoarding and heating, and work may proceed much more slowly.

for a more effective maintenance program and a more effective workflow procedure when addressing unplanned outages.

Procurement of Spares, Warehousing, and Maintenance Workshop

Procurement of spares and warehousing of critical spares are integral parts of any continuous operation system. Procurement and warehousing should generally fall under the control of the maintenance team in the operations organization. Processes must be developed to ensure all spares are properly catalogued and managed to functionally support the needs of an operating facility. Storage bins must be appropriately labeled to promote ease of access and reorder of consumable items. High-value critical spares must be properly secured. A *just-in-time* reorder system promotes cost saving and reduces the need for costly storage space.

An effective warehouse system must cater to the needs of the owner's operating and maintenance workforce while maintaining inventory control on use of warehoused items. In small- to medium-scale operations, only the operating staff may be required to work on a continuous (24-hour, 7-day-per-week) operating cycle. Access to the warehouse is therefore very important to the operating staff so that minor maintenance activities can be performed by operating personnel.

Clearly defined and agreed-to rules between both the maintenance and the operating organizations are required to ensure smooth operations of a warehouse. A procurement system that is timely and caters to the needs of both the maintenance and the operating teams within the operations organization is essential for warehousing success. A card access system provides access to all personnel during periods when the maintenance team is away from the facility. A jointly agreed-to set of rules helps in creating more harmonious relationships between maintenance and operating personnel in a facility. Joint ownership of the process through consultation when rules are developed helps in adoption of and adherence to agreed-on rules regardless of how difficult they may be.

A maintenance workshop is also required to provide maintenance workers a safe and comfortable workspace to perform work. While most maintenance work will occur in the field in the form of preventive maintenance, repairs must be done in a workshop to avoid injuries and maximize labor productivity and efficiency. A workshop must be properly equipped with tools, cleaning areas, workbenches, and power sources. Rules governing workshop etiquette must be established and adherence maintained. Proper signage is also required to create a safe workplace. Good housekeeping, reflected in clean floors free of tripping and fall hazards and a well-organized workspace, not only minimizes workshop injuries but also improves worker productivity.

Contractors and Service Providers

Identification and selection of service providers require careful attention. An effective contracting process depends on the owner's ability to manage the various stages of the contract life cycle, which include

1. Contracting strategy
2. Screening, evaluation, and selection
3. Contract approval
4. Execution and monitoring
5. Termination and closeout

Where contractors and service providers are in high demand, procurement of external services can be difficult and places greater responsibility on the owner to ensure quality service is provided. Where a facility may be a small business operating in areas dominated by larger business, special incentives must be available to ensure contractors and service providers are available to meet the needs of the facility. Maintenance planning must factor the availability of contracting services for planning and scheduling major activities such as plant or facility turnarounds.

The Contracting Strategy

Having an effective contracting strategy is an important requirement for success in project execution. A contracting strategy may be dependent on the geographic location and availability, skills, and ability of contractors and service providers, reliability, prior working experiences, safety records and performance of contractor or service provider, and resourcing ability of the contractor or service provider. A request for proposal (RFP) may work best for providing a single, one-time service, whereas ongoing service agreements may work best when both the owner and contractor or service provider may have prior joint working experience or for providing long-term ongoing support and services to the owner. In all cases, care and attention are required when deciding which contractors should be invited to bid on work for the organization. Adequate data collection from the contractor is required to make informed decisions during the selection process.

Screening, Evaluation, and Selection

Many databases and tools are available to assist in the screening, evaluation, and selection of contractors. While price is a determining factor in selection, other important considerations may include reliability, availability, asset control and skills capabilities, ability to provide specialized services, safety records and performance, quality of work, and ability to work with the owner in a collaborative way.

Careful screening, evaluation, and selection are required to minimize cost associated with a nonfunctional working relationship with a contractor. A weighted average approach to selection may be required by which weights are assigned to each variable important to the owner for selection. Contractors are scored against these variables. Table 6.2 provides a simplified screening, evaluation, and selection weighted average process.

While selection may not always be dependent on the highest weighted average score, the weighted average scoring process assists in the selection process in an objective and auditable manner. As shown in Table 6.2, contractor/service provider 5 has the best chance of selection based on the scoring method with a weighted average score of 200.

Contract Approval

The contract approval stage requires careful understanding of the scope of work to be completed by the contractor or service provider. Joint discussions between the owner and contractor or service provider are required to ensure full alignment on deliverables, cost structures, methods of payment, and other information pertinent to the contracted work. In all cases, the scope of work must be carefully written into the contractual agreement or work agreement before work can begin. An inspection and verification of the contractor's or service provider's equipment and machinery and workshop may be necessary to confirm ability to perform the assigned work.

Execution and Monitoring

A prejob meeting is required to reconfirm the owner's expectations with the contractor or service provider. Expectations concerning schedule, quality of work, safety culture, and other working agreements are reaffirmed with the contractor or service provider. Consistent with the concept of "what gets measured gets done," jointly agreed measures must be used to monitor progress and for payment and corrective actions as required.

A process for ensuring contractors are properly oriented at the facility is required. Knowledge of site rules, owner's codes of practices, and access to information required to perform work safely is necessary for the contractor or

TABLE 6.2

Contractor Selection Using a Weighted Average Scoring Process

Selection Variables	Variable Weight	Contractor/Service Provider 1		Contractor/Service Provider 2		Contractor/Service Provider 3		Contractor/Service Provider 4		Contractor/Service Provider 5	
		Ranking	Score	Ranking	Score	Ranking	Score	Ranking	Score	Ranking	Score
Price	10	4	40	5	50	5	50	3	30	5	50
Reliability	8	5	40	3	24	5	40	5	40	5	40
Safety performance	9	5	45	5	45	5	45	4	36	5	45
Reputation	6	5	30	3	18	5	30	3	18	5	30
Previous working experience	7	5	35	3	21	4	28	3	21	5	35
Total			**190**		**158**		**193**		**145**		**200**

Note: Assessment guidelines are as follows:
1. Contractor/service provider are assessed based on criteria (selection variables) deemed important to the owner.
2. Weights applied based on the relative importance of that variable to the owner.
3. Weight × Ranking = Contractor/service provider score for that variable.
4. Contractor/service provider final score = Σ scores for each variable.
5. Contractor/service provider with the highest score more likely to be selected than those with lower scores.

service provider to perform work safely. Joint contractor or service provider inspection will be required to assess work completion and progress so that appropriate payments can be made for work completed.

Termination and Closeout

Contract termination and closeout is an area often neglected by many organizations. Termination and closeout (assuming the work was completed as requested and as per contract) refers to opportunities for capturing learnings from the work that was just completed—identification of activities that went well and those that did not and why. In addition, both the contractor and the owner may have useful feedback and contributions to enhance performance should a similar job present itself in the future. Furthermore, both parties have a wonderful opportunity to further the benefits of relationship management through the closeout process.

On most occasions the termination and closeout is poorly accomplished because people are excited when the job is completed and owners become focused on the next contractor or the next activity while the contractor is focused on the next job. In addition, there is a bit of ambiguity as to who should complete this closeout exercise. Closeout in my view is a joint exercise between the field leadership responsible for supervising work performed by the contractor and procurement or supply chain management (SCM) personnel who are responsible for administering the contract. A meeting among the three stakeholder groups coordinated by the SCM organization provides an ideal opportunity for capturing learnings that can be beneficial to other parts of the organization and that can be easily shared.

Support Services: Human Resources, Accounting, Communications, Information Technology, Employee Assistance Programs

Adequate support systems are required to ensure a motivated and healthy workforce. Failure to pay personnel on schedule can create undue stress on workers, who depend on their wages and salaries to meet personal commitments. Systems to ensure workers are paid accurate wages on time are absolutely essential. Some tolerance is allowed by new employees with start-up operations, but for most employees tolerance is low for late payment of wages and salaries.

Systems to ensure proper cash flow management, budgeting, and allocation of assigned charges are necessary for proper management of the facility and for identifying areas of focus in cash flow management. Systems for communication and access to information systems must also be developed and shared with the workforce so that work can be accomplished properly. Support networks for employee health and safety must also be set up consistent with regulatory requirements (health and safety committees) and employee support networks.

Relationship Management

As with all other business areas, relationship management is an essential process for new projects. Leaders must actively develop processes for ensuring that trusting business relationships are developed with regulators, vendors, suppliers, and contractors or service providers. Demonstrated trustworthy, honest, fair, and consistent behaviors are required by leaders to build credibility and relationships with stakeholders. The ability to do the right thing at all times helps in developing trust with stakeholders. Saying what will be done and doing what is promised helps in developing credibility and business relationships.

Critical Ranking of Operating Areas and Operating Personnel Allocation

For effective and continuous operation of each operating area of the facility, a process must be developed to ensure operating personnel and resources are effectively allocated to ensure all operating areas are adequately staffed based on the strengths, skills and capabilities, and experience of the available operating workforce.

Considering a typical steam-assisted gravity drainage (SAGD) operation, an operating facility will typically be broken up into five operating areas as shown in Table 6.3. By applying a weighted average scoring system and assigning a risk score to each operating area, a criticality score can be assigned to each operating area. A criticality score for each operating area is essential for assisting the operations organization in allocating operating personnel based on personnel strengths and weaknesses. In this way, experienced and skilled personnel can be assigned to operating areas with high criticality scores to enhance the steady-state operations of that area.

Shown in Table 6.3, both the control room and water treatment operations received similar criticality scores of 194 and 193, respectively. As a consequence, when assigning operating personnel to these areas, leadership must

TABLE 6.3

Criticality Ranking of Operating Areas for Personnel Assignment

Criticality Variable	Variable Weight	Operating Areas									
		Control Room		Oil Separation		Water Treatment		Steam Production		Well Production	
		Risk Ranking	Score	Risk Ranking	Score	Risk Ranking	Score	Risk Ranking	Score	Risk Ranking	Score
Potential to impact personnel safety and operating efficiency	10	5	50	5	50	5	50	5	50	5	50
Potential to significantly impact production	9	5	45	4	36	5	45	4	36	4	36
Technical knowledge of operating guidelines and specifications	8	5	40	4	32	5	40	3	24	1	8
Mechanical and operational complexity	7	5	35	3	21	4	28	3	21	2	14
Workload	6	4	24	3	18	5	30	3	18	2	12
Total			194		157		193		149		120

Note: Operating areas with the highest risk scores will require personnel of the highest levels of competence evidenced by training, skills, and experience. Assessment guidelines are as follows:

1. Criticality variables are determined based on potential to impact steady-state operation and production.
2. Weights are applied based on overall impact to sustained operations and personnel safety.
3. Weight × Ranking = Criticality score for that variable for the specific area.
4. Area score = Σ scores for each variable.
5. Areas are risk ranked in descending order. The areas with the highest score have the greatest risk.

TABLE 6.4

Operating Personnel Assignments Based on Strengths and Weaknesses of Workers

		Worker Selection for Control Room										
		Tom		Mary		James		John		Henry		Jane	
Selection Criteria	Variable Weight	Rating	Score	Rating	Score	Rating	Score	Rating	Score	Rating	Score	Rating	Score
Prior experience in control room operations (minimum requirements 5 years)	10	4	40	4	40	4	40	5	50	3	30	5	50
Communication skills	9	4	36	4	36	5	45	5	45	5	45	5	45
Works well under pressure	8	5	40	5	40	5	40	3	24	4	32	4	32
Computer skills	7	4	28	5	35	4	28	5	35	4	28	4	28
Troubleshooting skills	6	5	30	3	18	4	24	5	30	5	30	4	24
Ability to multitask	5	4	20	3	15	4	20	3	15	2	10	3	15
Total			194		184		197		199		175		194

Note: Rankings are as follows:

1. Selection criteria are determined based on potential to have an impact on steady-state operation.
2. Weights applied based on importance of the particular skills in control room operations.
3. Weight × Rating = Selectivity score for that variable for the specific employee.
4. Employee score = Σ scores for each variable.
5. Selection criteria are rated in descending order. The highest criteria rating indicates greatest importance to the role.

carefully consider the skills and experience of the entire operations work-force to determine which personnel will be assigned work in these areas.

Table 6.4 provides a similar assessment of the skills, experience, and capabilities of personnel for allocation to work areas. Allocating workers in a greenfield situation involves critical thought, transparency, and involvement of frontline leaders (shift leaders). In a continuous operation shift environment, shifts must be adequately staffed so that strengths and weaknesses are balanced across all shifts. Failure to create this balance leaves shift leaders uncomfortable, unhappy, and often complaining.

Imbalances in shift strengths and capabilities lead to different performance across shifts. This also results in adverse impacts on equipment and machinery, leading to more frequent outages and breakdowns. Another potential outcome of imbalances in operating strengths and weaknesses across shifts is the creation of a breeding ground for poor performance and lack of effort.

The method for selecting and allocating workers across each operating area is defensible and transparent to both shift leaders and employees. In the example provided in Table 6.4, employees with scores highlighted in bold would be selected for a four–shift rotation system. Workers who are not selected form a pool of potential succession planning resources and should be developed as opportunities present themselves. They can then be used for filling roles in the control room when opportunities are available. Transparency assists in adoption and demonstrates fairness in the process. A similar process should be used for all critical roles in the organization when multiple potential candidates aspire to them.

Key to the success of this selection method is the involvement of shift leads, who may have had the opportunity to work with and observe the work behaviors, strengths, weaknesses, and attitudes of workers. The process is repeated for each operating area based on criticality ranking of the operating area and the selection criteria defined for each operating area.

7

Systems Readiness

Introduction

This chapter deals primarily with the mechanical hardware associated with the facility. *System readiness* refers to the mechanical completeness of all operating systems of the facility. When constructing large commercial facilities or operating plants, a shrewd project leader will break down the work primarily into areas and into systems associated with that area. For a typical steam-assisted gravity drainage (SAGD) facility, work may be broken into two major areas (production pads and processing facility) with a possible third area defined as the interconnecting pipelines and road infrastructure. Breaking down the work into such work areas allows construction leaders to define all systems within these operating areas and systematically drive them to completion. Table 7.1 provides a sample breakdown of systems to be completed and turned over to the operations organization by the construction and commissioning organizations.

Before any system is turned over to the operations organization, the construction and commissioning organizations will typically be responsible for ensuring that the system meets the following criteria:

1. Mechanical completeness
2. Design quality and operating standards and adherence to a management of change (MOC) process
3. Controls tested within design limits
4. Regulatory compliance
5. Safety systems installed and functional
6. Live systems communication notices
7. System walkdown, deficiency identification, and deficiency resolution
8. System turnover documentation completed and adequately packaged

TABLE 7.1

Typical System Identification for an Industrial SAGD Facility

Project	Areas	Systems
30,000-Kbpd steam-assisted gravity drainage facility	Main facility	System 1: Instrument and industrial air System 2: Power distribution System 3: Boiler feed water ⋮ System *n*: Dry oil to market
	Production pads	System 1: Instrument and industrial air System 2: Instrument power distribution System 3: Steam distribution and injection ⋮ System *n*: Produced fluids to main facility
	Interconnecting pipelines and road infrastructure	System 1: Pipe rack supports System 2: Interconnecting main steam line System 3: Interconnecting produced fluids main line ⋮ System *n*: Road networks

Note: kbpd = thousands of barrels per day.

Involving operating personnel in some of these activities provides the best learning opportunities for the operations organizations. Inspection of the internals of vessels, installation of critical equipment, and commissioning and testing of critical equipment enables the operations organization to benefit tremendously in conceptualizing how the system functions once all flanges and manways are closed. The use of commissioning procedures and adherence to codes of practice (COPs) and standard operating procedures (SOPs) are critical to minimizing injuries, failures, and damage to equipment, machinery, and the environment. If necessary, commissioning procedures and SOPs must be appropriately updated and reissued consistent with the field experience during commissioning and start-up. It cannot be overemphasized that the benefits of having personnel from the operations organization working alongside the commissioning and construction organizations during the late stages of system readiness are invaluable.

Mechanical Completeness

Mechanical completeness of a system refers to the complete installation of vessels, machinery, and all related hardware associated with the particular system. Mechanical completeness of the system refers to readiness of the system to be turned over to the commissioning organization so that the system can be commissioned and started up safely and consistent with the designed production capabilities of the system. For example, a mechanically complete instrument and industrial air system refers to complete installation of primary and backup compressors, surge drums, distribution headers with pressure release and safety protections systems. All flanges are closed and bolted, and mechanical valves are properly bolted or welded at connection points.

Mechanical completion must be consistent with blueprint drawings and equipment designs. Any variations from design specifications must follow an effective MOC process. The MOC process must properly analyze and consider the impact of design changes on the introduction of new hazards and ability to meet design process specifications. Failure to do so can result in disastrous consequences. Mechanical completion also includes adherence to quality standards by testing welds and joints for engineering quality standards. X-rays and magnetic particle inspections are tests that assist in establishing quality assurance on weld joints. Hydrotesting helps in establishing the integrity of the entire piping and interconnecting infrastructure consistent with design limits. Vessel integrity is often confirmed at the vendors' shops before being shipped to the facility.

Design Quality and Operating Standards and Adherence to an MOC Process

Prior to turning over any system to the operations organization, the construction and commissioning organizations must ensure the system conforms to design quality and operating standards. Line flushes, steam blows, suction strainers on pumps, and visual inspections are methods used by the commissioning organization to remove unwanted materials from the inside of piping and for protecting equipment from damage during the commissioning stage of execution. Documentation and commissioning checks must confirm the integrity and operating limits of the system through testing.

The benefits of an effective MOC process cannot be overemphasized for ensuring design quality and operating standards. Experience has confirmed that disastrous consequences can result from failure to perform an effective MOC. For example, removal of seemingly inconsequential drains on the low section of a steam header can result in water hammer with pipe rupture and explosions during the warm-up of the steam header. MOCs address the concerns of process safety hazards and risks introduced into a system and mitigate these hazards and risks to manageable levels through engineering controls, procedures, and as the last resort personal protective equipment (PPE). During project execution, the project leadership team must have an MOC process that works well. The right representation from required disciplines is an absolute during the change review process.

Controls Tested within Design Limits

During the system readiness process, the commissioning organization must verify that all control systems work properly and in accordance with design specifications. The commissioning and construction organizations must ensure that the distributed controls system works properly to allow operating personnel to maintain control of the system from a central point or the control room. All control valves must be checked with the control system to ensure that it opens when the control room command opens and vice versa. Software programming and control logic must be checked, double-checked, and verified as functioning correctly.

Regulatory Compliance

Before any system is turned over to the operations organization, the construction and commissioning organizations must ensure that the system meets full regulatory compliance. This may include, but is not limited to, regulators governing the environment, operating of pressure vessels, natural resource management, and social and community affairs. Consultation with these regulating agencies if necessary and approval must be sought and documented before proceeding to start-up or turnover to operations occurs.

Safety Systems Installed and Functional

Prior to system readiness and turnover to the operations organization, all design safety systems must be tested and available for use. Personnel must also be trained and qualified in the use of the safety systems. Safety shutdown systems, fire suppressant systems, eye wash stations, safety showers, and other portable safety systems all must be checked and confirmed functional before system readiness and turnover are possible.

Live Systems Communication Notices

Before commissioning work begins, personnel must be aware that the system is being livened up and that process fluids will be running through the particular system. Notices as shown in Figure 3.5 should be posted across the system to inform personnel that process hazards are now introduced into the system and additional attention is required when working with this system. Involving the operations organization in releasing this type of communication notice helps to ensure that the safety of all personnel is maintained. System owners must also ensure that interconnection points on system boundaries are properly isolated and tagged to warn personnel who may work on these interconnecting boundaries of the hazards within the pipes and the need to consult system owners before work can be done at those boundary interconnecting points.

System Walkdown, Deficiency Identification, and Deficiency Resolution

Often it is not possible for the construction and commissioning organizations to complete all the work on a system before it is turned over to the operations organization. Once the operating integrity and safety standards of the system are met, the system may be turned over to the operations organization with an agreement to complete all outstanding work within a reasonable and agreed-on period.

When purchasing a newly constructed home, for example, before the home owner takes ownership of the home from a builder, the owner and the builder's representative walk through the home identifying and listing all outstanding work that must be completed by the builder before the home purchase agreement can be deemed completed. A similar process occurs when a system is to be turned over to the operations organization. Turnover sequence and requirements are best demonstrated in Figure 7.1.

Representatives of the construction, commissioning, and operations organization are required to walkdown the system to ensure compliance with design requirements. Representation from all three organizations is an effective means of identifying deficiencies that may belong to the respective construction or commissioning organization before the system is turned over to the operations organization.

The joint construction, commissioning and operations system walkdown and deficiency identification process (commonly referred to as punchlisting) is best demonstrated in Table 7.2. As shown in Table 7.2, deficiencies can be classified as A or B deficiencies.

- A: Deficiencies represent those deficiencies that have the potential to present an unsafe operating condition or adversely affect operating performance.

- B: Deficiencies represent those deficiencies that may cause an unsafe condition that can be mitigated against in the short term. A B deficiency does not impede production of the system.

When assigning completion dates, care must be taken to ensure A deficiencies are completed consistent with milestone dates. In the case of the milestones established in Figure 4.3, March 24 was identified as M1 (milestone 1), turnover of the first system. If the instrument and industrial air systems

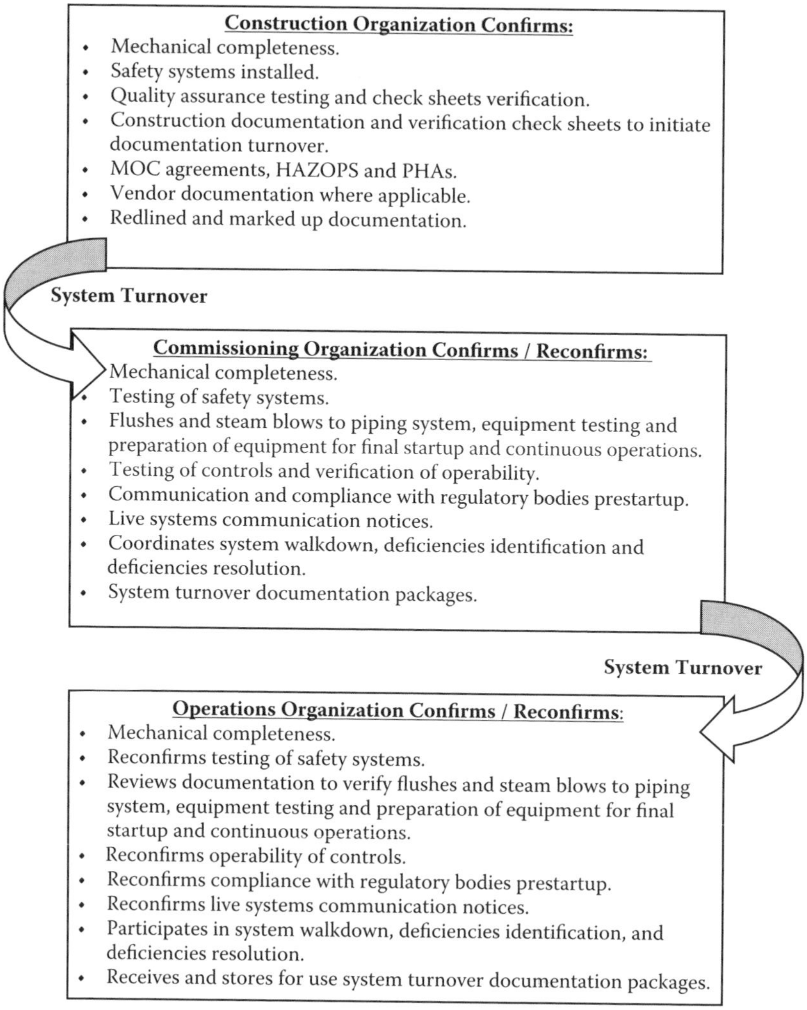

Construction Organization Confirms:
- Mechanical completeness.
- Safety systems installed.
- Quality assurance testing and check sheets verification.
- Construction documentation and verification check sheets to initiate documentation turnover.
- MOC agreements, HAZOPS and PHAs.
- Vendor documentation where applicable.
- Redlined and marked up documentation.

System Turnover

Commissioning Organization Confirms / Reconfirms:
- Mechanical completeness.
- Testing of safety systems.
- Flushes and steam blows to piping system, equipment testing and preparation of equipment for final startup and continuous operations.
- Testing of controls and verification of operability.
- Communication and compliance with regulatory bodies prestartup.
- Live systems communication notices.
- Coordinates system walkdown, deficiencies identification and deficiencies resolution.
- System turnover documentation packages.

System Turnover

Operations Organization Confirms / Reconfirms:
- Mechanical completeness.
- Reconfirms testing of safety systems.
- Reviews documentation to verify flushes and steam blows to piping system, equipment testing and preparation of equipment for final startup and continuous operations.
- Reconfirms operability of controls.
- Reconfirms compliance with regulatory bodies prestartup.
- Reconfirms live systems communication notices.
- Participates in system walkdown, deficiencies identification, and deficiencies resolution.
- Receives and stores for use system turnover documentation packages.

FIGURE 7.1
System turnover sequence and requirements. HAZOPS, hazards and operability studies; PHA, process hazards analysis.

TABLE 7.2

Typical Deficiency (Punch List) for a System Walkdown

Project Title: Company A SAGD Project
System 1: Instrument and industrial air system
Date: March 16 (Year)

Deficiency Type		Summary Description of Deficiency	Remedial Actions	Responsible Organization	Proposed Completion Date	Date Completed	By Whom – Block Letters and Signature
A	B						
X		Compressor K-2 trips on high vibration after running for approximately 15 minutes on 60% load.	To be filled in when remedial actions have been completed.	Commissioning	March 18		
	X	Water leg on surge drum heat traced but not insulated.		Construction	March 30		
	X	Very noisy in compressor operating station. Check noise level and reconfirm compliance with regulatory standards.		Construction/ commissioning	April 10		
	X	Inadequate labeling of process pressures in system piping.		Construction	April 10		

Representation Approval: _____ Construction _____ Commissioning _____ Commissioning _____ Operations _____ were among the M1 systems, there will be very little time to effect remedial actions on the A deficiency identified.

In the example provided, the A deficiency requires immediate repairs before the system can be turned over. All B deficiencies can be mitigated against without affecting the health and safety of personnel until permanent repairs can be completed. Production from the system is also not affected by these deficiencies. Heat tracing can be temporarily wrapped with insulation, noise concerns can be mitigated against with procedures and PPE (earplugs and earmuffs), and signage can be improved with notices and temporary signs.

System Turnover Documentation Completed and Adequately Packaged

The system turnover process is a formal transition of ownership from one organization to another during the execution stage of a project. Appendix 4 provides a detailed example of the responsibilities and turnover process. The process is validated using a common form that shows details and authorities who may be involved in the turnover of systems. The form is used to verify the turnover transactions between the construction organization and the commissioning organization in the first instance at the date the commissioning organization is able to begin functional testing and commissioning activities. The form also verifies the turnover transactions between the construction organization and the operations organization.

Details of checks and tests to be performed by the construction, commissioning, and operations organizations are shown in Appendix 5. Once the commissioning organization has completed its functional testing and is ready to turn over the fully operational system to the operations organization, the formal process is again followed, with the relevant section of the form used to validate this transaction. Figure 7.2 demonstrates both transitions of ownership through to the operations organization—the final owner.

System turnover documentation is quite possibly one of the most important activities in system readiness during project execution. Documentation is important to establish that the owner received a facility consistent with design requirements from the contractor selected to build the facility. Document turnover signifies a formal handoff of the system to the operations organization from the construction and commissioning organizations.

Binders containing all relevant documentation regarding vendor specifications, testing to confirm operating specifications, quality assurance checks and verifications, and all relevant documentation for continuous operation of the system are passed on to the operations organization. Duplicate copies

System Identification Number: _____1_____

System Name: _____**Instrument Air System**_____

Turnover: Construction Organization to Commissioning Organization

The system indicated above has reached essential construction completion and is ready for system testing and commissioning activities to commence. This system is in compliance with design requirements. A system punch list has been prepared and reviewed for completeness by the signatories. All construction testing data is attached to this form.

Complete System Turnover   **Partial Turnover**

_____   _____

Construction Leader _____

_____   _____

Commissioning Leader _____

Turnover: Commissioning Organization to Operations Organization

The system indicated above has been successfully tested and is ready for normal operation. The updated system punch list has been reviewed and no outstanding A-Deficiencies have been identified capable of affecting the health and safety of personnel or normal operations and production consistent with design standards. All construction and commissioning testing data are attached to this form.

Turnover by _____   Date _____

Commissioning Leader _____

Concurrence by _____   Date _____

Construction Leader _____

Acceptance By

Operations Leader   Date
(Owner's Representative) _____   _____

FIGURE 7.2
Formal system turnover form. (© Suncor Energy Inc. With permission.)

are turned over to the owner so that one copy can be stored at the site for site use and the other copy may be stored at a safe location for backup availability and insurance needs in the event of unforeseen events to the facility. Appendix 6 provides a sample list of typical turnover documentation required by a facility.

8

Stakeholder Management

Introduction

Stakeholder relations are an essential requirement during project execution and must be effectively managed. Generally, a specialized and expert group of project support personnel may be tasked with managing stakeholder relations. Key to success in stakeholder relations is to ensure that all stakeholders are properly identified and that consultation, collaboration, and communication are done in a proactive, clear, and transparent manner. Honest, factual, and consistent communication behaviors are absolutely required.

Project leaders must ensure that stakeholder interests and expectations of each stakeholder group are clearly defined and sufficient efforts are made to properly manage these expectations. Figure 8.1 provides a list of internal and external stakeholder groups.

Internal and External Stakeholder Groups

The expectation of each stakeholder group must be understood by the project leadership team and strategies developed to manage these expectations on a stakeholder group basis. Stakeholders must be prioritized in terms of relative importance to the project and resources committed to ensuring adequate and timely consultation, communication, and involvement.

Project leaders must not underestimate the strength of various stakeholder groups. For example, in the absence of proper consultation and involvement, aboriginal groups can render a halt to project execution if their expectations and interests are not properly managed. Similarly, regulatory bodies can halt execution if not properly consulted or informed. Nongovernmental organizations (NGOs) such as environmental groups may also halt a project in the absence of sufficient communication, consultation, and mitigation actions to address their needs.

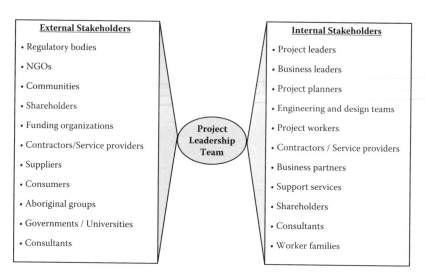

FIGURE 8.1
Stakeholder groups.

As far as stakeholder management is concerned, the focus should be primarily on avoiding bad news to stakeholders. When bad news must be provided to any stakeholder group, this must be done with full empathy and caring behavior and a demonstration that adequate efforts were made to avoid the undesirable situation. Involving expert communication personnel is important to properly present bad news such that undesirable consequences for delivering bad news can be mitigated against.

Stakeholder Communications

In *all cases*, honesty, timeliness, transparency, and communicating from the heart with demonstrated empathy and care for the stakeholder group may reduce the level of adverse consequences. Such communication behaviors help in developing trust between the owner (and project team) and the stakeholder group.

To assist junior leaders in responding to stakeholder concerns, a good practice is to locate posters of regulatory stakeholders in prominent locations that are accessible in the event there is need to contact the stakeholder. In most instances, regulators will provide to the project reportable requirements and limits and are often prepared to conduct necessary training when applicable. Taking advantage of such offers symbolizes a willingness to do the right thing by project leaders. In addition, project leaders required for interacting with the media must be identified and trained to respond to media questions such that they are able to properly communicate in the event that the need may arise for doing so.

It is absolutely imperative that project leaders maintain continuous vigilance via a 24/7 phone number and by having a stakeholder management process to ensure proper feedback to the affected individual or associations when operating within culturally or environmentally sensitive regions or areas. Recall Syncrude Canada Limited's incident in early 2008 when approximately 500 migrating ducks died when they landed in a tailings pond in the northern city of Fort McMurray. According to Austin (2008), "An anonymous tip alerted officials about 500 birds were in the pond" (par. 3). Despite the small number of birds involved in this unfortunate incident relative to its population, the scrutiny and negative publicity associated with the event has had an impact on Syncrude both directly (fines) and indirectly (image concerns and unintended scrutiny). More important, when sensitive areas are affected, such situations can lead to increased scrutiny from ENGOs (environmental nongovernmental organizations), NGOs, and the public about the safety of other sites. Industry peers will also be affected by such negative publicity and will often not look kindly on such incidents. The key therefore is to maintain continuous vigilance over such sensitive situations to avoid negative publicity and scrutiny.

It is difficult to communicate anything positive to stakeholders in such an incident other than the preventive actions taken and a reassurance that such incidents will not recur. Early estimates of direct fines from Alberta Environment for this incident are estimated at approximately $1.0 million (Haggett 2008). This fine translates to approximately $2,000 per duck. Proactive responses to sensitive areas through early recognition, continuous vigilance, and training of all personnel in proactive responses can avoid similar situations. Preventive measures enable organizations to generate tremendous goodwill with many stakeholders both internal and external to the organization and help in creating a socially responsible corporate image. In the example provided, timely installation of noise cannons may have potentially prevented the public relations nightmare and enforceable corrective actions to which the organization may have exposed itself. In addition, had an employee of Syncrude, as opposed to an anonymous caller, reported this incident to regulators with appropriate corrective actions, the fallout would have likely been less damaging to the organization.

Stakeholder Interest and Expectation Map

Although not exhaustive, Figure 8.2 provides typical stakeholder interests and expectations for four different stakeholder groups. Similar detailing will be required for each stakeholder group. The ability to prioritize among

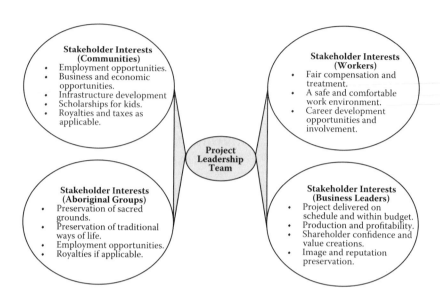

FIGURE 8.2
Typical stakeholder interests and expectations.

these interests and communicate consistent messages across all stakeholder groups enhances the success of stakeholder relations and management. The availability of specialized support services to deal with these issues and guide the project leadership team in an advisory capacity reduces distractions from the project execution focus.

9

Managing the Budgets

Introduction

The budget management process is a secondary component of the overall project execution process. Similar to any business process, a prudent fiscal management process is required to operate within budgets. The focus of cost control for project leaders is generally on the variable cost streams, which include the following:

1. Wages and salaries
2. Managing employee turnover
3. Waste management and rework
4. Competitive pricing structures
5. Cost escalation from lagging labor productivity
6. Cost associated with project schedule slippages
7. Managing project scope
8. Treating expenses associated with safety as a benefit as opposed to a cost

Wages and Salaries

Wages and salaries are an important element of the overall project budget. The ideal situation is to pay workers fairly based on industry standards. Ideally, in a steady-state operation an employer should seek to position wages and salaries marginally above industry average, allowing noncash benefits to promote market advantages for wages and salaries to the organization.

During the execution stage of a project, the goal of the project leader must be to attract the best and brightest. Competitive wages are necessary to ensure that quality workers are attracted and retained. In the absence of a competitive compensation structure, the project may fail to attract the experience and technical excellence required to support high levels of productivity required for scheduled delivery of the project. In addition, employee turnover can be high, with consequential cost and schedule implications for the project.

Managing Employee Turnover

Shaw, Gupta, and Delery (2005) advised that turnover adversely affects financial and workforce performance "through its impact on skills and motivation" (p. 3). Phani (2006) and Price, Kiekbusch, and Theis (2007) reiterated the cost and performance issues related to worker turnover in different industries. Research suggested that employee turnover affects the health and stability of an organization, and that turnover represented a loss of "productivity and its employees acquired knowledge" (Institute of Management and Administration 2007, 1) from the organization.

Associated with worker turnover are cost increases arising "from knowledge loss and retraining, new employee orientation and training, as well as time to productivity" (Schwieters and Harper 2007, 78). Research also pointed to the fact that employee "replacement cost can be as much as 150% of the departing person's salary" (Branch 1998, par. 3). Green (2004) estimated the cost of employee turnover to be the equivalent of employee pay for 1 year. When considered in its entirety, I advised that the real cost of employee turnover includes costs associated with hiring, lost productivity from turnover of employees, lost productivity from learning impact of replacement hire, training costs, and "the time consumed by co-workers in training the new worker" (Lutchman 2008, p. 7).

According to Meisinger (2006), in today's knowledge-based economy, "the experience, skills and creativity of people is the most important factor in creating and sustaining an organization's competitive advantage" (p. 10). Green (2004) advocated preventing turnover through effective hiring practices for organizational and cultural fit. As a consequence, organizations must do more to retain their personnel. According to Fields et al. (2005), factors leading to turnover can be categorized as shown in Table 9.1. Such categorization "can help leaders better identify some root causes of turnover and develop policies, practices, and organizational behavior strategies that can lead to the long-term reduction and control of worker turnover" (Lutchman 2008, p. 41).

We have often heard the expression "Workers leave bosses ... they do not leave jobs." I am a strong supporter of this notion having seen similar situations in the workplace many times. Building on the recommendations

TABLE 9.1

Factors Influencing Worker Turnover

Worker Turnover Category	Specific Factors
Employee demographics	Include attributes such as employee "age, education, gender, tenure and family responsibilities" (Fields et al. 2005, 63)
Job-specific characteristics	Include attributes such as "security, skill variety and autonomy, job stress and job satisfaction" (Fields et al. 2005, p. 63)
Organizational behavior characteristics	Include attributes such as leadership and "supervision, pay and benefits and current performance ratings" (Fields et al. 2005, 63).
External environment characteristics	Include attributes such as available job opportunities, unemployment rates (Fields et al. 2005), and status of the economy or specific industry

Source: Lutchman, C. Leadership impact on turnover among power engineers in the oil sands of Alberta. Dissertation, University of Phoenix, 2008, p. 42.

of Green (2004) of hiring for organizational and cultural fit, I strongly recommend project leaders invest the time and money required to develop the leadership skills of their supporting leadership teams. Weak leadership capabilities can be extremely costly to a project during the execution phase from employee turnover and accompanying productivity decreases from the workforce. The situation is worsened when turnover leads to schedule delays.

Project leaders must ensure the competency of their supporting leadership teams and focus on developing leadership skills within their teams; these skills include engagement and motivation, team building, individual consideration of workers, training and development of the workforce, cultural awareness, and genuine care for workers. When workers are exposed to these types of leadership behaviors, worker satisfaction is generally high. Overwhelmingly, the research identified that poor job satisfaction is a cause for worker turnover (Conklin and Desselle 2007; Daly and Dee 2006; Lambert, Hogan, and Barton 2001; Zellars et al. 2005). The goal of the project leader and the leadership team during project execution, therefore, is to maximize worker satisfaction during work.

Waste Management and Rework

One of the key goals of the project leadership team during project execution is to get the job done right the first time and to do so with minimal waste of resources. Careful planning and prework before a job is undertaken are necessary to facilitate this process. Adequate prejob instructions and work supervision help in getting the job right on the first attempt. The use of proper tools and equipment also helps in ensuring the quality of the job. Needless to say, the importance of competent personnel cannot be overstated for getting the job done right the first time and in an efficient manner.

To avoid rework, an effective quality standards and quality assurance program is required. Project leaders must resist the impulse to hasten workers during supervision. Rather, the focus should be on supporting the needs of the worker and enabling the worker to work efficiently and safely. Rushing through work encourages workers to take shortcuts, which generally result in shoddy work and often the need for rework. Ultimately, rework results in increased cost during project execution and must be avoided at all times.

Competitive Pricing Structures

Successful cost control during project execution depends on an effective supply chain management process. An effective sourcing system is required to secure competitive pricing structure to support project execution needs. The focus on competitive pricing during project execution must be primarily on consumable items. For large industrial projects, capital items would have been ordered well in advance of execution, so there are few opportunities to influence cost at this point. Nevertheless, changing business environments may provide opportunities to renegotiate costs associated with large and expensive capital items. Just-in-time delivery leads to avoided warehousing costs; however, delays in delivery can result in schedule delays. A careful balance is therefore required to manage risks associated with delayed deliveries relative to project schedule delays.

Cost Escalation from Lagging Labor Productivity

Fatigue, overexertion, and insufficient off time and rest days from work can lead to a slow but marked reduction in labor productivity. Project leaders must establish the right balance between the length of a workday and the amount of consecutive workdays during which workers are expected to perform before a break is allowed. This balance may vary for the three organizations, with the construction organization working the most consecutive numbers of days, generally in the range of 10–14 days at work and 4 days off. The operations and commissioning organizations may work according to a 5-day, 40-hour work cycle, conforming to a normal work week. Often, extended work is required by both these organizations throughout project execution to support the needs of the construction organization.

As project execution continues and as shown in Figure 4.4, when systems turnover among the organizations begins to occur, both the commissioning and the operations organization may adapt the same work cycles as the construction organization. At this stage, project leadership vigilance over productivity, burnout, and employee safety is paramount to sustained cost and scheduled deliverables.

Indeed, as the end of project execution nears, when workers look into the future and envisage no jobs to transition into on completion of the existing project, there is a tendency to reduce productivity to extend the duration of current employment. Such productivity impacts have the potential not only to increase labor cost but also to create a large holding cost associated with the use of equipment and machinery such as cranes and lifting equipment. In addition, there is increased potential for rework and scope creep at this stage. Leadership must therefore develop creative means for maintaining high labor productivity and keeping workers focused on the tasks at hand as opposed to the distractions beyond completion of the project.

Cost Associated with Project Schedule Slippages

The greatest potential for cost escalation comes from project schedule slippages. Schedule slippages lead to additional labor cost and large holding cost associated with equipment and machinery. This holding cost can be worsened based on the prevailing business environment. In a heated market economy with full employment, both labor and holding costs are high, and each day of slippage can result in hundreds of thousands, if not millions, of dollars in cost overruns.

Depending on the nature of the delay, some of this cost can be passed on to suppliers and other stakeholders. However, when delays are related to people and process readiness, this cost must be borne by the owner and must be reflected as cost overruns to the project.

The readiness process provides an early warning mechanism for identifying and proactively addressing potential delays in people and system readiness. When used properly and with the right criteria selection and focus, owner-related delays and cost can be avoided or minimized. The trick is to ensure all relevant people and process readiness criteria are identified, resourced properly, and stewarded with the right quality control checks and balances.

Managing Project Scope

Understanding the competing priorities of the various organizations, as discussed in Chapter 2, helps in managing scope and avoiding scope growth or scope creep. Recall that the construction organization wants to build the project (systems) and move on to the next challenge; the commissioning organization wants to demonstrate that it works, and then move on to the next challenge (generally following the construction organization). The operations organization wants to have the best-functioning and -operating facility for the long-term future.

With this understanding, the scope of smaller jobs during project execution can be properly managed. Maintaining an effective management of change (MOC) process allows the three stakeholder groups to effect desired scope changes in a managed way to avoid compromising the integrity and operability of any system. An Energy Resources Conservation Board (ERCB) of Canada investigation report (2008) reviewed the circumstances surrounding the catastrophic water hammer failure of a main steam header line on a new construction project. Reconstruction and costs, although not quantified, were substantial to the project. The project underwent approximately 1 year of delay before it was granted approval to be restarted. The investigation report pointed to concerns associated with the MOC process. Concerns were also found with procedures applied at the project as well as people and process readiness concerns relating to training and standard operating procedures (SOPs). Their records management processes were also found to be deficient.

Treating Expenses Associated with Safety as a Benefit as Opposed to a Cost

The safety performance during project execution is critical to sustained high productivity and performance. It must be noted that during this stage of a project, the risk exposures of personnel are extremely high. The situation is made even riskier when the work site is inhabited by multiple organizations with large numbers of personnel sharing a limited work site. Weak and poor safety performance during project execution can result in lowered worker morale, worker turnover and flight, lowered worker productivity, and increased costs associated with managing incidents.

Project leaders must ensure that adequate safety and loss management resources personnel are available among all organizations to mitigate against the undesirable consequences of safety management system failures. Emphasis should be placed on both leading and lagging safety indicators to protect the safety of the workforce and assets. Project leaders must recognize that costs incurred in managing the safety of the workforce during project execution generate a tremendous return on the investment, although it is difficult to quantify.

Money is best spent on proactively managing leading indicators since this type of investment not only reduces incidents but also demonstrates to the workforce that leadership cares for their well-being, and this can enhance the motivation of the workforce.

Leading indicators may include:

1. Safety training courses administered to the workforce
2. Safety audits performed at the work site
3. Frontline leadership visits by senior leadership personnel
4. Number of safety alerts issued
5. Number of near misses experienced
6. Number of safety stand-down town hall meetings conducted

Lagging indicators may include:

1. Total recordable injury frequencies
2. Total disabling injury frequencies
3. Total first aid administered
4. Loss time incidents sustained at the work site

Safety during project execution is the responsibility of every worker. Workers must be encouraged to stop unsafe work and to look out for each other. The most vulnerable groups during project execution may be inexperienced workers and immigrant workers, for whom language and cultural differences may influence communication skills and safety behaviors. Proactive training and addressing these issues can result in tremendous cost savings during project execution. *The safety of workers during project execution comes before cost and schedule concerns.*

10

Situ-Transformational Leadership Behaviors: A Model for the Future

Introduction

Leadership during project execution by far exceeds any other variable in determining the outcome of a project as far as schedule and budgets are concerned. As previously stated, "leadership is about creating an organizational environment that encourages worker creativity, innovation, proactivity, responsibility and excellence" (Lutchman 2008, p. 19). Leadership is about creating a compelling sense of direction for followers while motivating them to performance levels that will not occur in the absence of the leader. Indeed, Blanchard (2008), one of the premier experts on leadership, advised that there is no single best leadership style for leading any workforce. However, the powers of the principles of the situational leadership and transformational leadership behaviors working together during project execution cannot be underestimated in building loyalty and motivating workers to higher levels of performance.

In most instances, *workers leave leaders as opposed to leaving jobs*. The cost of replacing workers who leave the job is high. Organizations may incur costs from turnover that range from 33% to 150% of the annual salary of the replacement hire (Agrusa and Lema 2007; Branch 1998). When considered in its entirety, direct tangible replacement cost may include recruitment and selection cost, reduced productivity associated with new worker learning, training and development expenditures, and other incentives offered to attract and retain a new hire (Branch 1998). On the other hand, intangible costs such as knowledge, experience, skills, and creativity, which contribute to competitive advantage, can also be high in a mature workforce (Meisinger 2006). Ashworth (2006) made it clear that workers' knowledge, skills, competence, and capabilities represent valuable assets of an organization. As a consequence, this asset must also be leveraged for optimal performance and value maximization for the organization and its stakeholders. To do so, satisfaction must be high and workers motivated to perform.

The importance of understanding leadership behaviors for promoting a more satisfying work environment cannot be understated in work environments characterized by full employment and when difficulties are encountered in attracting new and skilled workers. The recommendations offered in this chapter provide opportunities for employers to reduce turnover-related costs and to improve individual worker and overall organizational performance.

Leadership must be addressed on two fronts during project execution: leadership at the front line and senior leadership or strategic-level leaders. At both levels, expectations of followers are ultimately reduced to the same denominator: satisfaction. Leaders must create a work environment in which followers are able to derive a level of job satisfaction that will encourage them to perform at sustained high levels of productivity and build their loyalty to the organization.

The Frontline Leader

The first point of contact between workers and the rest of the organization is through frontline leaders and supervisors. The way frontline leaders and supervisors treat workers will influence workers' motivation, commitment, performance, productivity, and loyalty. The onus is on employers, therefore, to properly equip frontline supervisors and leaders with the skills and capabilities to positively influence the behaviors of workers. Frontline leaders must be equipped with leadership skills and behaviors that create a more satisfying work environment for all workers so that workers may do their jobs safely at high levels of performance and so that organizations can retain these workers for longer periods and maximize value-driven performance during their employment tenure.

Worker satisfaction is a leading indicator of employee turnover (Allen, Drevs, and Rube 1999; Cooper-Hakim and Viswesvaran 2005; Price, Kiekbusch, and Theis 2007; Slattery and Selvarajan 2005; Trevor 2001). Furthermore, research has linked leadership to worker satisfaction and performance (Madlock 2008; Sharbrough, Simmons, and Cantrill 2006). When workers are provided with a more satisfying work environment, they are less likely to leave, they are more motivated, and they produce more (Bass 1990; Hoffman 2007; Owens 2006). During project execution, worker continuity is essential for keeping projects on schedule and reducing cost. As a consequence, the leadership behaviors demonstrated by our frontline leaders can significantly influence schedule and cost outcomes during project execution.

While there are many variables for creating a satisfying work environment, leadership is one variable that can be easily managed through training and development so that leaders can positively influence the behaviors of followers.

Leaders who can create a more satisfying work environment for followers can ultimately help in motivating workers to higher levels of productivity, reducing worker turnover, and contributing to the competitive advantage of the organization. As a consequence, leadership styles and behaviors that motivate workers can be promoted in the workplace by appropriate training and development opportunities for leaders. Leadership is highly influential in determining the quality of the immediate work environment in which workers perform and is therefore critical in determining the level of job satisfaction derived by workers (Price, Kiekbusch, and Theis 2007).

All organizations are generally characterized during the project execution stage by a workforce of varying skills and capabilities. Therefore, leaders must be able to adapt leadership behaviors to meet the requirements of the individual workers who make up the workforce. Ken Blanchard's Situational Leadership® II provides an effective model for leading and developing workers through the various stages of worker maturity. Blanchard (2004) pointed out that when employers invested time and leadership to improve the maturity of workers, such employers experienced strong gains and productivity improvements. Table 10.1 shows the various stages of a worker's development and leadership focus during each stage of the worker's development.

According to Blanchard (2008), Situational Leadership II requires varying levels of directing and supporting behaviors from leaders based on the maturity of the worker. Blanchard suggested that leaders apply four sets of behaviors relative to worker maturity. Figure 10.1 demonstrates the leadership behavior requirements of each stage of the worker development to maximize satisfaction and worker productivity.

During the execution stage, strong situational leadership skills and transformational leadership behaviors are required for leading an effective workforce. Shown in Figure 10.1 is the application of the principles of situational leadership and transformational leadership behaviors designed to improve the skills, capabilities, and motivation of the worker (situ-transformational leadership behaviors). Figure 10.1 shows the transformational leadership behaviors required by supervisors at each stage of worker development. These leadership behaviors help in developing the worker from incompetent and noncommittal to competent and fully committed.

Ultimately, the goal of any leader is to create a low-maintenance worker. The principles of situational leadership help in doing so. Successful transition of workers from high maintenance to low maintenance requires that frontline leaders be adequately trained to lead the frontline workforce. Such training must include training in both situational leadership principles and transformational leadership behaviors. Discussed in Chapter 2, Figure 2.3 defines the skill requirements for the frontline leader, and organizations must provide training opportunities to frontline leaders during project execution to ensure these skills are developed in the frontline leader to successfully manage and lead the frontline workforce.

TABLE 10.1

Leadership Behaviors with Maturity Status of Worker

Leadership Behaviors	Directing/Supporting Relationship	Worker Maturity
Directing (stage 1)	High directing/low supporting	Immature: Low competence and commitment; leadership focus on 1. Telling the worker where, when, and how to do assigned work 2. Key requirements of structure, decision-making control, and supervision 3. Primarily one-way communication
Supporting (stage 2)	High directing/high supporting	Immature: Growing competence; weak commitment; leadership focus on 1. Building confidence and willingness to do assigned work 2. Retaining decision making 3. Promoting two-way communications and discussions
Coaching (stage 3)	High supporting/low directing	Mature: Competent; variable commitment; leadership focus on 1. Building confidence and motivation; promoting involvement 2. Allowing day-to-day decision making 3. Active listening and two-way communications and discussions
Delegating (stage 4)	Low supporting/low directing	Mature: Strong competence; strong commitment; leadership focus on 1. Promoting autonomy, decision making, and empowerment 2. Collaborating on goal setting 3. Delegating responsibilities

Source: Adapted from Blanchard, K., *Leadership Excellence*, 25(5), 19–19.

It is worth noting that the leadership behaviors and traits of the frontline leaders are the outcome of the training provided to frontline leaders and the culture of the organization. Simply put, the *organizational culture is a reflection of the way we do business,* and frontline leaders are required to develop and encourage behaviors consistent with the values demonstrated by the organization. Training provided by the organization is intended to support these values. Figure 10.2 demonstrates the relationship among the organizational culture, organizational values, and leadership traits.

During project execution, leaders must seek to support the organizational values by demonstrating leadership behaviors consistent with these values. The organizational values of *trustworthiness and respectfulness* require demonstrating leadership behaviors of fair treatment to all workers, engaging and motivating workers, embracing diversity, and building trust and relationships with stakeholders. For the inexperienced leader, developing these skills

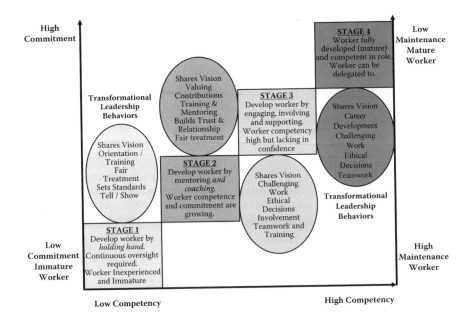

FIGURE 10.1
Situ-transformational leadership.

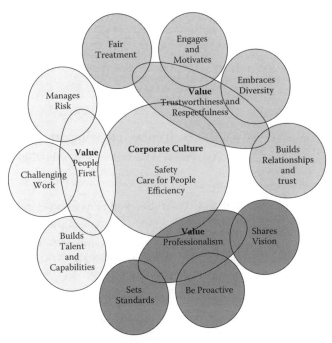

FIGURE 10.2
Corporate culture, values, and leadership traits.

and behaviors can be difficult. For example, expressive reactions and behaviors may fail to build trust and relationships. A valued friend and colleague, Don Clague, vice president in situ operations, Suncor Energy Incorporated, once said to me, "The way a leader responds to bad news determines how much bad news he or she will receive in the future" (personal communication 2008). What he meant by this was that expressive reactions lead to followers avoiding the responsibility of bearing bad news. He advised, "When there is no more bad news, this does not mean all is well, instead, expressive leaders will hear about the bad news much later on." In the project execution work environment, leaders must develop behaviors that encourage workers to bring forward bad news so that timely corrective actions can be applied. The earlier project leaders know of bad news, the faster they may be able to respond to correct this situation.

From a leadership perspective, leaders must develop behaviors that promote trust with the workforce, and the way a leader responds to the workforce will determine the level of trust that is developed. Training helps in developing the leadership skills of frontline leaders and must be a high priority for all organizations prior to and during the execution phase of a project.

Success with workers at the front line requires that frontline leaders demonstrate behaviors that set standards, are proactive in responses, build trust and relationships, engage and motivate the worker, develop self and workers, and provide the ability to leverage existing tools of the organization. At the front line, these behaviors must be reflected in a way that allows frontline workers to understand and respond to the frontline leader.

Set Standards

Setting standards is demonstrated by the following frontline leadership behaviors according to Suncor Energy (D. Clague, personal communication 2008):

- Frontline leaders ensure all workers (employees and contractors) have all required safety training before being assigned work.
- They ensure all workers have access to relevant SOPs [standard operating procedures], COPs [codes of practice], critical practices, and regulatory training.
- They verify worker competency by observing workers at work.
- They ensure adequate supervision is provided to all employees and based on the maturity of the worker.
- They promote involvement, collaboration, and feedback to workers on work-related performance.

- They ensure close monitoring and control of safety-sensitive work situations in the workplace.
- They seek help and input when work-related situations are difficult and larger than expected.
- They ensure all hazards are identified and all workers are aware of them.*

Proactive in Responses

According to Suncor Energy, behaviors that may indicate to workers that a frontline leader is proactive include the following (personal communication 2008):

- Frontline leaders will take immediate actions to stop unsafe work. They will also empower workers to shutdown/intercede in all unsafe work.
- Frontline leaders will ensure proper permitting process and worksite walkaround (before, during, and after) work.
- They will ensure all work is preceded by toolbox and pre-job meetings that promote worker engagement and where all workers feel free to voice concerns about assigned work.
- They will ensure proper scoping and risk mitigation plans are in place and reviewed with workers before a job begins.
- Frontline leaders will enable all workers to document incidents, near misses, hazards, and mitigations in a formalized process.*

Build Trust and Relationships

According to Suncor Energy, behaviors that may indicate trustworthiness and team-building capabilities of a frontline leader include the following (personal communication 2008):

- Frontline leaders promote a collaborative and consultative work environment with all workers for the safe completion of all work.
- Frontline leaders recognize the value of honesty and demonstrate care for all workers in a credible way. They are known for walking the walk regardless of how difficult the decision may be.
- They ensure that the safety of all workers is placed above all else while work is being performed.

---

* Copyright Suncor Energy Inc.; approval from Suncor Energy Inc. is required to reproduce these data.

- They demonstrate genuine care and empathy for workers and embrace diversity in the workforce.
- They promote transparency and are not afraid to acknowledge having made an error or mistake.*

Engage and Motivate the Worker

Behaviors that may indicate engagement and motivation of workers by a frontline leader include the following (personal communication 2008):

- Workers are consulted and encouraged to provide input in how work is to be completed. Where input is not considered, feedback to workers is provided in a fair and impartial manner on the reasons for pursuing alternative options.
- Genuine care and empathy for workers.
- Treating all workers fairly and with respect.
- Rewards are shared among workers who perform well. Workers have opportunities to determine who gets rewarded from whom does not within work teams.*

Develop Self and Workers

According to Suncor Energy, behaviors that may indicate a frontline leader's ability to develop self and others may include the following (personal communication 2008):

- Frontline leaders maintain continuous vigilance on the capabilities of the workforce under his/her supervision. Addressing gaps in skills and training of workers are priorities for the frontline leader.
- Frontline leaders seek necessary resources to ensure all workers are trained with and strive for balance between training while maintaining continuous operations.
- Frontline leaders continually seek to ensure competency of the workforce as opposed to checkmark compliance where critical training is required.*

For both employee and contractor frontline supervisors, the ability to leverage existing tools of the organizations is important to overall success. Existing tools of the organization that may support frontline leaders may include the following:

---

* Copyright Suncor Energy Inc.; approval from Suncor Energy Inc. is required to reproduce these data.

1. Access to information such as SOPs and codes of practices required for the safe execution and completion of the work.
2. Access to the organization's incident management systems such that near misses and findings from incident investigation reviews can be shared among workers.
3. Use of the organizations computerized maintenance management system to ensure maintenance and preventative maintenance work can be properly scheduled and addressed.

In my view, frontline leaders must be able to proactively address the needs of the workforce to develop credibility and gain trust from workers. As a consequence, the organization has a responsibility to prepare frontline leaders so that they may proactively respond to the needs of workers. The principles of situational leadership allow frontline leaders to recognize the maturity state of the worker and to respond to worker needs with transformational leadership behaviors as shown in Figure 10.1.

Senior Leadership (Strategic Leader)

At senior levels in the various project organizations, leadership must demonstrate primarily transformational leadership behaviors. According to Bass (1990, 53) transformational leaders encourage followers to "transcend their own self interest for the good of the group, organization or society; to consider their long term needs to develop themselves rather than their needs of the moment; and to become more aware of what is really important." Transformational leaders focus on developing the workforce and creating a work environment in which workers feel a sense of belonging, and they are treated fairly; they are motivated and are provided intellectually stimulating and challenging work. Transformational leaders create strong teams by leveraging the abilities of experienced workers while developing less-experienced workers. In the author's view, the paradigm of the transformational leader is to motivate workers to do more than is expected of them by leveraging their creativity and excellence.

The author also suggests that such leaders will create an organizational environment that encourages creativity, innovation, pro-activity, responsibility and excellence. They often possess moral authority derived from trustworthiness, competence, sense of fairness, sincerity of purpose, and personality. Transformational leaders are also trustworthy and ethical in decision making. Trust is earned by leaders, and it results from consistent demonstrated behaviors of doing the right thing at all times. In addition, transformational leaders have the unique ability to communicate and share the organizational vision. They promote involvement, consultation, and participation. Transformational leaders demonstrate high levels of emotional and cultural intelligence and will successfully lead in volatile, fast-paced, ambiguous work environments that are characteristic of project execution environments.

Transformational leaders have been credited with the ability to promote teamwork and develop strong teams. Research has shown that workers led by transformational leaders exhibit high job satisfaction (Korkmaz 2007). Hoffman (2007) established that leadership influences employee satisfaction and ultimately employee retention and loyalty. Project execution relies almost entirely on teamwork. Failure of any work group to deliver on milestone deliverables will affect the performance of other work groups, and as such, communication, collaboration, and teamwork are required to succeed. Senior project leaders must foster and support teamwork among the construction, commissioning, and operations organizations.

Motivating Employees

Wren (1994) discussed the early pioneering work of Elton Mayo on how to motivate workers. Mayo pointed out that wages and job characteristics are strong motivators for workers. Fair and competitive wages during project execution motivate workers to remain employed. However, job characteristics are more important in determining the level of productivity of workers during project execution. Challenging and intellectually stimulating work encourages workers to be committed and to produce more. Ramlall (2004) suggested motivation affects productivity and employee turnover behaviors. During project execution, all leaders must strive to limit turnover since the cost associated with turnover can be both in terms of both lost productivity and lost knowledge.

Gentry (2006) suggested leadership behaviors can influence undesirable worker behaviors such as absenteeism, tardiness, and turnover. According to Gentry, leaders may consider behaviors that reflect fair treatment of all workers, genuine empathy, teamwork, and involvement in workplace-related decisions to address some of these undesirable worker behaviors. During project execution, leaders should therefore communicate with followers in a manner that builds trust within the workforce. *Saying what you will do and doing what you say is so very important in building trust.* Workers are motivated to emulate the behaviors of leaders who make ethical and trustworthy decisions. All workers seek fair and unbiased treatment, and leaders who provide this gain both the trust and the respect of all workers.

Employee motivation can be high when leaders are trustworthy, teamwork is dominant in the workplace, workers are treated fairly and individually, and leaders act with empathy when communicating with workers. Leaders must say what they intend to do and do what they promise to do. Ramlall

(2004) identified the following factors that leaders can address to influence worker motivation in the workplace:

1. The personal needs of the employee: May include but are not limited to training, personal protective equipment, or simple advice on a personal issue. Leaders must be able to recognize the body language of workers who may have unresolved personal issues and respond to them with the empathy required.

2. The work environment characteristics: Generally, commercial and industrial project environments are characterized by difficult working conditions. Unpaved roads, extreme temperatures, outdoor work, and supplemental lighting can all influence the productivity and motivation of workers.

3. The responsibilities and duties of the worker: When placed in supervisory roles, worker motivation can be affected either positively or negatively. If unprepared for the role, negative consequences may be generated. If equipped to do the job and if seen as a promotion, workers may respond positively.

4. The level of supervision provided to the worker: Immature workers will necessarily require more supervision and guidance relative to experienced workers, who may require less supervision. Micromanagement, constantly telling a worker how to do the job without consideration for the worker's maturity, can lead to experienced workers *quitting* and immature workers *staying* because they need delegation.

5. The extent of worker effort required to perform assigned tasks: Labor intensiveness versus skill intensiveness influences worker motivation.

6. The employee's perception of organizational fairness and equity: Perceived unfair and biased treatment is a precursor to worker turnover and level of effort generated from workers. Workers tend to equate work with pay, and when faced with possible unfair treatment they will find creative means to bring about equity. Shirking on the job, engaging in absenteeism, arriving late to work, and providing poor performance are methods used to level the playing field.

7. Career development and advancement opportunities: Most workers have an unspoken personal development plan for the workplace. They will be motivated to perform at high levels if career development opportunities are aligned with their personal plans.

According to Ramlall (2004), employers should seek to respond with empathy to the personal needs and values of the employee. They should also seek to create work environments that are respectful, inclusive, and productive (Ramlall 2004; Wren 2004).

Employee Training and Development

Workplace training can result in greater "job satisfaction, organizational commitment and turnover cognition" (Owens 2006, 166). Satisfaction and commitment will ultimately translate into higher organizational performance (Price, Kiekbusch, and Theis 2007). Little (2006) found intent to turnover was lower in organizations that looked after the training needs of workers relative to those that did not. In a research study, Little found 41% intent to turnover within the year in organizations with poor focus on training relative to 12% where training was made a priority by employers. People readiness is a critical component for being ready to transition from one milestone to another during project execution. People readiness requires continuous training and retraining of workers to maintain high levels of productivity. When workers feel that they are unqualified for a job and they are unable to learn quickly enough in the role to undertake the task, they will leave.

According to Finegold, Mohrman, and Spreitzer (2002), when training permitted the development of new skills, satisfaction and loyalty were higher among younger and inexperienced technical workers relative to older workers. Finegold, Mohrman, and Spreitzer found that across all age groups satisfaction with continuous skills development showed a strong relationship to organizational commitment and loyalty. Clearly, therefore, emphasis on training and development for all workers provides opportunities for organizations to increase not only the technical competence of workers but also their commitment and loyalty to the employer. Training for all personnel serves to reduce turnover during project execution and is an absolute requirement for all personnel. Project leaders must recognize trained workers as an asset and not view training as a cost. A training needs analysis must be conducted for all employees, and proactive responses to closing these gaps are required.

In my experience, when workers were trained properly, their confidence levels were increased in the tasks they were assigned to perform. Training not only enabled skills and capabilities but also motivated workers to perform assigned tasks quickly, safely, and efficiently. Training also creates a line of sight between the current state of the worker and future career development roles, leading to high motivation to perform in the workplace.

Involvement and Participation in Workplace Decisions

Collaboration and involvement in workplace decisions influence worker loyalty and employee retention (Agrusa and Lema 2007). During project execution, leaders are continually seeking ways and means to improve tasks and the way we do work. Involving experienced workers in workplace decisions is critical in the continuous improvement process. According to RoSPA (2006), involving workers in organizational safety can result in improved organizational safety performance. Collaboration and involvement promote ownership and buy-in to work activities and a greater willingness by workers to get the job done right. As a consequence, leaders who involve workers in workplace decisions and carefully consider inputs from their followers will likely see more support for new activities in the workplace.

Involvement does not necessarily mean accepting the input from followers; instead, leaders should give consideration to input with follow-up communication on the justification for exclusion or inclusion of such inputs. Workers are extremely appreciative of feedback and information that explains why their input may not have been accepted and will often lend support for the accepted decisions once they receive proper communication. While leaders may find it difficult to allocate time to consider the input of all participants and to provide feedback when input is not considered, the benefits of doing so in building trust, commitment, and motivation are high and should not go untapped.

Research has shown that employee involvement in decision making at appropriate levels influences turnover decisions by workers (Guthrie 2001; Wilson, Cable, and Peel 1990). According to Riordan, Vandenberg, and Richardson (2005), participation in workspace (immediate workplace) decisions led workers to feel more valued by their employers. Riordan and coworkers found worker perception of involvement in decision making influenced organizational effectiveness. According to Lockwood (2007), organizations that involved workers in business decisions benefited from a more diligent, loyal, and motivated workforce. Lockwood also found that when employers failed to involve workers in business decisions, they were more likely to experience higher employee turnover increases and lower productivity. Involving workers in workplace decisions is a step in the right direction for increasing worker motivation, productivity, and loyalty. Such worker behaviors are critical during project execution.

Teamwork

Teamwork during project execution can be encouraged by leaders promoting social events and family activities. Promotion of cross-organizational social events encourages integration of the entire project execution workforce. Golf tournaments, fishing expeditions, and family picnic days can generate significant team building across all organizational groups during project execution. Furthermore, the importance of training provided to senior leaders on how to build and sustain strong teams cannot be understated.

Teamwork (peer and leader support) is a leading contributor to worker loyalty (Pisarski et al. 2006; Thompson 2002). Teamwork that promotes coworker encouragement, support, advice, approachability, listening skills, and emotional intelligence led to more satisfying work environments for workers (Pisarski et al. 2006). Teamwork and team building depend on collaboration and sharing among workers (Rosen and Callaly 2005; Silén-Lipponen, Tossavainen, Turunen and Smith 2004). Teamwork and a sense of belonging by workers when worker input is considered with open-mindedness can enhance worker performance and loyalty (Glisson and James 2002; Pisarski et al. (2006). Transformational leadership behaviors support teamwork and the creation of strong teams. Strategies for teamwork and team building during project execution are absolute requirements.

Harris, Kacmar, and Witt (2005) found poor-quality leadership support led to poor job satisfaction and commitment among workers. According to Mansell, Brough, and Cole (2006), leadership support was a strong predictor of intent to turnover by workers. Leadership support positively influenced the workspace of workers and their productivity and should be an important consideration for all leaders. According to Smither (2003), when workers were treated individually (leaders avoid lumping into a collective group), workers were more motivated and were less likely to leave an employer. Knowing your workers by their first names, taking time away from the office to be visible and available to workers, and responding to their needs with genuine interest are strong motivators for all employees, regardless of their level in the organization.

Smither (2003) recommended responding to the needs of every worker individually with genuine empathy to each worker's personal and career needs. Genuine empathy and honest, open communication that is free from repercussions serve to motivate workers and contribute to a trusting transformational work culture in which workers want to stay and perform (Ramlall 2004; Smither 2003). Workers will leave when they feel they cannot communicate with their supervisors or leaders and their opinions are not valued (Thornton 2001).

Leaders must seek to create a satisfying work environment for all workers (Hopkins and Weathington 2006; Price, Kiekbusch, and Theis 2007). Creating such a work environment depends on the leader's ability to provide challenging work, promote involvement and participation among workers, provide equal access to training opportunities, promote teamwork and team building, and gain the trust of workers (Desselle 2005; Price, Kiekbusch, and Theis 2007).

Trust in leadership and the organization is a key outcome from the demonstrated actions of the leader and employer. Communication methods, channels, and behaviors adopted by leaders influence the levels of trust workers will place with employers (Hemdi and Nasurdin 2006; Hopkins and Weathington 2006). Trust is high when employers demonstrate genuine concern and empathy for workers (Hemdi and Nasurdin (2006). Trust in leadership is earned when leaders demonstrate competence, emotional intelligence, integrity, and ethics in decision making (Davis et al. 2000). In the project execution environment in which the safety of personnel depends on the decisions made by project leaders, such leaders must earn the trust of all workers. Care for people and cultural intelligence in a multicultural and diverse project execution environment help in maintaining trust in leadership.

Critical Thinking Ability

A leader's ability to critically analyze issues and to eliminate personal biases adds to organizational success during project execution. Critical thinking skills are learned behaviors, and all senior leaders in a project execution work environment must be able to think critically. The ability to probe deeply and to understand root causes of issues both contribute to overall success during project execution.

Facione (2006) pointed out that the characteristics of a critical thinker include inquisitiveness, a desire to be well informed, self-confidence, open-mindedness, trustfulness of reason, flexibility, fair-mindedness, honesty about personal biases and prejudices, openness to opinions of others, and lack of fear to change course when reflection suggests it is necessary to do so. Harris (1998) suggested that leaders must be competent in recognizing and removing obstacles to critical thought that may impair the conclusion and results of work if not addressed in the decision-making process. According to Harris, obstacles may include behaviors such as prejudices and stereo-typing, defensiveness, language interpretation difficulties, emotion fixation, channel vision, learned helplessness, and psychological blocks. Critical thought promotes creativity in the workplace and is absolutely desirable during project execution.

Michalko (2000) advised that "creativity deviates from past experiences and procedures" (p. 18), thereby creating room for thinking outside the box. How can we enhance our capability to think critically? Leaders who develop habits that recognize and address prejudices, stereotyping, and logical fallacies enhance their critical thinking capability and skill development. Leaders who perform in project execution roles should be taught skills to improve critical thinking skills for maximum performance during this stage of the project.

Critical thinking is the *art* of forming right conclusions from a body of information provided on any particular subject. The specific application of the word *art* suggests that critical thinking is a learned behavior that encourages the critical thinker to clearly navigate through information with cognizance to avoid errors of fallacies, emotions, language influences, and other obstacles to critical thinking in developing conclusions from information. Leaders can inadvertently develop sacred cows and construct empires when placed in project execution roles when critical thinking skills are weak.

It must be noted that developing critical thinking skills starts with an acknowledgment of self-limitations. There must also be a genuine desire to improve capabilities in critical thought processes. Project execution leaders should be enlightened about the benefits of critical thinking. It is in response to perceived benefits of the critical thinking process that people are motivated to aspire to be better critical thinkers. The learning process starts with basic training in recognizing obstacles that get in the way of the right decision or conclusion. Figure 10.3 provides a simple exercise to demonstrate the

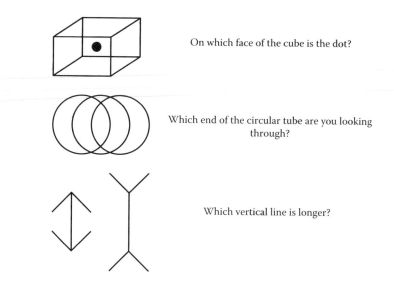

On which face of the cube is the dot?

Which end of the circular tube are you looking through?

Which vertical line is longer?

FIGURE 10.3
Critical thinking and perceptions. (Adapted from Results Based Interactions, *Leadership Skills,* Development Dimensions International Inc., MCMXCIX, Pittsburgh, PA, 2004.)

impact of perception on decision making. The examples demonstrate that some situations may have different answers based on individual perspective, as in the case of the dot on the face of the cube and the circular rings. The parallel line example highlights the importance that things may not always be what they appear to be.

11

Conclusion

There is no magic bullet for project execution that guarantees delivery within budget and on schedule. However, there are many methods used by project leaders for keeping projects as close as possible to planned budgets and schedules. The readiness method outlined in this book adds to the list of possible options available to project managers and leaders. The value of the recommendations proposed in this book is that they provide a set of practical tools and processes for early identification of issues and challenges that are likely to derail a project during the execution phase.

While there is little focus in this book on cost control for budget using sophisticated financial models and control software, the overriding premise is that once schedule is maintained, budgeted cost will be achieved (assuming cost structures were within the range of reasonable estimates during the project planning and scope definition stages). Of course, prudent management of financial resources consistent with good management is required. Focus on cost control should be on the controllable cost or variable cost streams to promote efficiency and a reduction of fixed-cost structures to minimum levels.

It is impractical to consider achieving budgeted cost for delivering a project when scope creep and scope increase are allowed in the project. Furthermore, the impact of market pressures that lead to escalating prices for commodities can lead to budgetary overruns. While scope creep and scope increases are manageable and additional costs associated with these events can be quantified and controlled, cost increases associated with market increases from labor and commodities pricing cannot be controlled and should not be included in the analysis when budgeted cost is compared with actual cost.

Success in project execution is dependent on the leadership skills of the project leadership team and having the right people and processes in place to support execution. Undoubtedly, the way we equip our personnel before they are allowed to function in frontline leadership roles in any of the organizations discussed in this book will ultimately determine commitment, motivation, and productivity of workers. Leaders must be trained properly and the owner's leadership expectations must be clearly understood before leaders are allowed to interact with employees at the front line.

Once qualified to lead at the front line, as representatives of the owners, standards must be set and expectations from all stakeholders clearly defined and communicated. The culture of the organization must be visible and dominant so that new employees and visitors can easily recognize and adapt to expected behaviors. The use of safety glasses, hard hats, protective gloves, earplugs or earmuffs, and fire-retardant clothing by all workers on a commercial oil and gas project define the safety standards and expectations for all workers. Signage and warnings and response to safety improvement suggestions by workers define leadership commitment to the safety culture of the organization.

Having a tool kit of practical templates, processes, and guides assists project leaders to better control the various events and processes for which they may be responsible. A tool kit also helps in supporting these leaders in the delivery of quality work on schedule and within budget. A tool kit is an effective means for reducing work effort required to complete assigned work in a work environment where leaders are already stretched to the limit.

The importance of fairness and transparency cannot be underestimated during project execution. Defensible processes as discussed in Chapter 6 that demonstrate fairness and equity when dealing with people are critical requirements for gaining the trust of the workforce. Every worker wants to know that he or she is treated fairly and, if possible, is involved in the decision-making process. Indeed, involvement promotes buy-in and acceptance. The full support of all stakeholders is required to ensure success during execution.

It is not uncommon to have projects completed on schedule and within budget and, when placed in the hands of the operating organizations, the projects fail to meet design production. In greenfield operations with new technology, such situations may arise as new knowledge is developed. However, there are occasions when design production may be inadvertently overstated, and cost trimming to gain corporate approval may lead to elimination of critical systems or equipment that may adversely affect the operation of the facility. In such situations, even though the project is executed such that it is on schedule and within budget, disclosed production targets may never be achieved with the existing design. An owner may pursue two options: The first is to accept the actual project in its current state and disclose revised performance targets. Adverse market response on an extended basis is likely from this option. Stakeholder confidence is eroded and may require major efforts to repair. The other option is to perform remedial work under the title of debottlenecking.

Debottlenecking is a process often undertaken to augment the performance of the project to meet disclosed design standards. Retrofitting cost can be huge since it involves working on a live facility, and additional cost is involved in cleaning and preparing work to meet work standards. The analogy I use for this type of situation is as follows: A buyer wants a Cadillac as a personal car, buys something that looks like a Cadillac, and spends the necessary cash to repair the look-alike to operate like a Cadillac. In this scenario, the buyer bears the cost of the first purchase, the additional cost of repairs, and loss of value from inability to use the vehicle while repairs are undertaken. Ultimately, this cost far exceeds the initial cost that would have been incurred if a Cadillac was purchased in the first instance. In such situations, project execution leaders have achieved the goal of *on schedule and within budget*.

Appendix 1: Sample Standard Operating Procedure*

Author:	Approved by:
Signature:	Signature:
Date:	Date:
Reviewed by:	
Date:	

Objective

To safely start up steam generator 04-SG-401C from

1. A cold start and

2. A warm/tripped restart

This procedure details steps to be followed after the steam generator has been filled and warmed up to design to a minimum of 135°C with circulating boiler feed water.

Hazards

- High-pressure and high-temperature boiler feed water.

- High-pressure and high-temperature steam.

- Rotating equipment.

- Hydrocarbon gases possibly containing H_2S.

- High-noise area.

- Hot surfaces.

Personal Protective Equipment (PPE) Requirements

- As per employer's standards on the named facility.

- Appropriate PPE, such as hearing protection, gloves, H_2S monitors, fire retardant coveralls, and hard hats.

Supporting Documents and Location

- Vendor documentation: Facility steam generators operation and maintenance manual

- P&ID A1-004-B-011: Steam generators auxiliary—water flows.
         A1-004-B-012: Steam generators auxiliary—air and gas flows—controls.

- Related SOPs on
    Feed water systems and pumps
    Filling steam generators
    Steam flow to high pressure (HP) separators
    Swinging from blowdown to main steam header

If any errors or omissions are found when using this procedure, please note such on this form and hand it directly to your immediate supervisor for incorporation of improvement suggestions.

Important

All four steam generators cannot be started up concurrently since the blowdown system used for startup has been sized to handle blowdown from a maximum of two steam generators at the same time. When more than two steam generators are required, two would be started up, and once these generators are placed into the main header, other required steam generators to a maximum of two can then be started. *Note:* While the first two are in blowdown mode, the other two can be in warm-up circulation mode.

Steps	Notes
A. Pre-Light-Off Checks	
1. CONFIRM that the combustion air preheater system 04-E-441C is in service.	Glycol circulating under temperature control through tubes.
2. CONFIRM that the forced draft fan motor 04-KM-401C is energized and ready for service.	Lockout devices removed and breaker energized. • Stack damper must be 100% OPEN to satisfy burner management system (BMS) light-off logic. • HVAC (heating, ventilating, air-conditioning) system in service supplying air to building to prevent creating negative pressure in the building.

3. CONFIRM these systems before attempting to fire steam generator: 04-SG-401C.	• Distributed control system (DCS) start-up sequence enabled (HS-1300).
	• Emergency local stop (HS-1300A) and remote DCS emergency shut down (ESD) (XX-1300A) satisfied.
	• Instrument air satisfied.
	• Stack damper limit (HZSO-1357) is satisfied by manually opening to 100%.
	• Pilot gas valves satisfied:
	  XZSC-1385A/B shutoff valves CLOSED
	  XZSC-1385C vent valve OPEN (double block and bleed).

4. CONFIRM main gas system and preignition interlocks (XX-1300C) satisfied.	• PI-1371B reads < 1 kPa. • FI-1370 reads 0 kPa. • XZSC-1375A/B shutoff valve CLOSED. • XZSC-1375C vent valve OPEN (double block and bleed). • HZSC-1376 supervisory valve CLOSED. • FZSC-1377 flow valve is at low-fire position.
5. CONFIRM produced gas system satisfied.	• XZSC-1365A/B shutoff valves CLOSED. • XZSO-1365C vent valve OPEN (double block and bleed).
6. CONFIRM water system satisfied.	LP feed water pumps 04-P-400A/B, and, HP feed water pumps 04-P-401A/B are ready for operation. See appropriate SOPs. • At least 1-HP pump running to satisfy DCS output XX-1300B to BMS.

Steps	Notes
7. CONFIRM XZSO-1301 water motor operated valve (MOV) is 100% OPEN.	Located downstream of 6-inch block valve on HP feed water line to generator.
8. CONFIRM feed water flow is established.	SEE: Appropriate SOPs for filling steam generators.
9. CONFIRM warm-up of the generator.	These limits are satisfied: • Water flow rates: FALL-1305/1310/1315/1320/1325 • Water flow differentials: FDAHH-1305/1310/1315/1320/1325 • Tube temperatures: TAHH-1308/1313/1318/1323/1328 • Steam outlet temperature TAHH-1331
10. CONFIRM readiness of forced draft fan.	FD fan switch set to AUTO: • KM-401C-AU-DCS system sees this status change. • FYI-1393 air inlet vane is driven to 0% open.
11. CONFIRM start of FD fan.	04-KM-401C local start or via DCS.
12. CONFIRM FD fan at full speed.	10 seconds minimum.
13. CONFIRM opening of air inlet vanes.	FYI-1393 to 100% OPEN.
14. CONFIRM FD fan at open limit.	FZSO-1393: May be restricted during winter (cold air conditions) due to possible fan motor overload.
B. Pilot Flame Light Off	
15. TURN local burner switch ON.	HS-1300-B.
16. CONFIRM ready-to-start light is ON.	Light XL-1300J on local panel.
17. PUSH the local START button HS-1300C.	• Must be depressed within 30 minutes to avoid FD fan shutdown caused by system being confused with a postpurge ending.
18. CONFIRM illumination of purge timer light.	Light XL-1300K: on a DCS-initiated 120-second purge timer sequence if following conditions satisfied: • FD fan on AUTO. • FI-1393 (air flow) > 75% flow rate. • PAHH-1351 comb. chamber trip not in trip condition. • Fuel permissive (pilot, main, and produced gases) satisfied. • CO below trip condition: AAHH-1353 at < 2,000 ppm.

Steps	Notes
19. CONFIRM extinguish of purge light and illumination of limits complete light.	XL-1300R upon purge completion. Limits complete output signal XX-1300E to BMS control and turns limits complete light XL-1300I ON.
20. WAIT out 45-second prestart timer sequence.	This mimics a 30-second flame relay penalty due to air inlet vane damper limit at 100%.
21. CONFIRM end of this sequence.	• DCS drives inlet vane damper FYI-1393 to low-fire position (10%). • Low-flame limit FZSC-1393 on inlet vane is satisfied. • Gas control valve FZSC-1377 satisfied for low fire.
22. CONFIRM illumination of low fire hold light XY-1300L.	• DCS sends low-limit-satisfied signal XY-1300F to flame relay. • Air flow FI-1393 > 25%. • DCS energizes pilot valve permissive XY-1300H. • DCS waits 2 seconds to allow low-fire permissive to be picked up by flame relay. • DCS sends 1-second start pulse XY-1300G.
23. CONFIRM illumination of igniter and pilot lights.	• DCS indicates igniter and pilot valve relays BS-1390 and XS-1385 respectively. • Turns on igniter XL-1300M and XL-1300N lights.
24. CONFIRM illumination of local BMS flame relay lights.	• DCS indicates pilot shutoff valves OPEN and vent valve CLOSED through limit switches XZSO-1385A/B and XZSC-1385C.
25. CONFIRM pilot ignition.	Igniter stays on for 5 seconds but flame detection period lasts 10 seconds. • If flame not detected, pilot gas valves will CLOSE. • Pilot vent valve will OPEN. • Generator needs to be repurged. The local flame relay in the BMS panel would need to be reset.
C. Main Flame Light Off	
26. CONFIRM gas-enabled sequence.	• OPENS shutoff valves XV-1375A/B. • CLOSES vent valve XV-1375C.
27. CONFIRM illumination of supervisory light.	• DCS detects limit switches state change on main gas valves. • Local panel light XL-1300P indicates OPEN supervisory valve. • Main fuel gas valves XL-1300-O OPEN.

Steps	Notes
28. OPEN manual supervisory valve.	• Operator has 5 minutes to open this valve HV-1376. Operator has timer-controlled 10 seconds to get it 100% OPEN once started; detected by HZSO-1376.
29. CONFIRM main flame established; pilot circuit shut down.	When main flame detected: • After 10 seconds, pilot permissive is removed. • Pilot shutoff valves CLOSE. • Vent valve OPENS. • Local Pilot ON light XV-1300N goes OFF.

D. Steam Generator Warm-up from Cold Start

30. LEAVE the steam generator on low fire.	• For a warm-up period of at least 5 hours (commissioning period recommendation).

E. Steam Generator to SERVICE After Cold Start-up or Warm/Tripped Startup

31. SWITCH firing valves to AUTO position.	Control Room Operator will • Switch air inlet vane control and main gas valve FV-1377 to AUTO and accept control of firing rate.
32. CONFIRM illumination of RUN light.	Low-fire hold light XL-1300L goes OFF.
33. INCREASE generator firing rate.	Control room operator will • Increase firing rate on generator while Maintaining feed water rate and generator back pressure control. Keep increasing firing rate slowly until 80% steam quality is achieved. • See SOPs for Warming line to HP separators. Swinging from blowdown to main steam header. • Required start-up flow: 37,000–50,000 kg/hour. • Steam is generated at 9,640 kPa, 309°C, and 80% quality.
34. INTRODUCE produced gas to burners (when available).	Control Room Operator will cause XY-1365A&B to open and XY-1365C to close (double block and bleed). PCV-1362 will maintain constant pressure to burners.

Steps	Notes
35. MONITOR generator operation.	The Area Operator will continue to monitor the generator and related equipment and area for 1. Leaks. 2. Flame pattern. 3. Interior hot spots (seen through furnace inspection ports). 4. FD fan bearing temperatures. 5. FD fan abnormal noises.
36. ENSURE chemical injection systems are in service.	• Sodium sulfite to suction of LP FW pumps. • Chelant to suction of LP FW pumps. • Filming amine downstream of HP separators.
37. ADJUST chemical injection rates as necessary.	As steam production is increased.
38. TEST steam samples.	• To determine steam quality of each pass of the steam generator. • Based on this information control room. • Operator will adjust flows as needed to maintain balanced flow through all passes.

Hazardous Task Control Analysis

Task Identified	Start-up Steam Generator 04-SG-401C Operating Area 4				
Potential Hazards	Hazard Rating	Score	Score	Hazard Rating	Potential Hazards
Temperature			2	4	Noise
>50°C	5	5		8	Inert gas
−20 to 50°C	1			1	Poor lighting
>−20°C	5			6	Ignition source
Pressures				3	Heavy Lifting
<50 kPa	1		3	5	Rotating/moving equipment
50 to 700 kPa	4			3	Using ladder
>700 kPa	7	7		3	Using scaffold
Chemicals					
Reactive	4			6	Trenching, excavating
Poisonous	10			3	Breaking system integrity
Corrosive	6			10	Confined space vessel entry
Electrical					
0–12 volts	1		3	6	Restricted access/egress (VENTS)
12–120 volts	3			3	Hoisting/rigging

Hazardous Task Control Analysis

Task Identified	Start-up Steam Generator 04-SG-401C Operating Area 4				
Potential Hazards	Hazard Rating	Score	Score	Hazard Rating	Potential Hazards
>120 volts	5	5		6	Crane work
Flammable (FUELS)	8	4		3	Overhead work above
Radiation	8			30	Cutting into a pipeline
Hazard Rating Score		21	8		
Total				21 + 8 = 29	

If combined total is ≥ 30, please proceed to the next page, Analysis and Control Form.

Hazard	Consequences	Existing Safeguards	Recommendations
>50 (oil temperature)	Exposed skin burned	Coveralls, gloves, etc.	Wear PPE and work carefully.
>−20°C (colder)	Frostbite	Coveralls, gloves, etc.	Wear PPE and work carefully.
12–120 volts	Electrical shock	Training, gloves, PPE, insulation, covers	Wear PPE and work carefully.
Gas pressure > 700 kPa	Embolism, eye injury, trauma	Training, PPE	Compressed air should never be directed at an individual's body.
Heavy lifting	Strains, sprains, fall	Training, PPE	Lifting assistance, training on lifting techniques.
Noise	Hearing loss	Training, PPE	Add a training statement on dangers of noise.

*Appendix 2: Sample Critical Practice**

<table>
<tr><td colspan="2" align="center">**Organization/Facility**</td></tr>
</table>

Critical Practice: Filling Acid Storage Tanks, 03-T-309 A/B

Reference Number: _____

Title: _____

Equipment ID: _____

Author: _____

Reviewed by: _____

Last Review Date: _____

Approved by: _____

Expiry Date: _____

Objective
- To safely offload bulk shipments of hydrochloric acid from a tanker into the acid storage tanks.

References/Supporting Documents
- Materials Safety Data Sheets (MSDSs) for 30–40% HYDROCHLORIC ACID. Located (identify location here).
- ACID/CAUSTIC OFFLOADING ERP 10-04. Located (identify location here).

Hazards
- CORROSIVE: May be fatal if inhaled or swallowed. May cause severe skin and eye burns and irritations.

Personal Protective Equipment Requirements
- As per organization/facility standards.
- Neoprene gloves, goggles, face shield, acid-proof suit, and steel toe rubber boots.
- Half mask with filters for acid vapors.
- PPE for all personnel as defined in the acid-caustic offloading emergency response procedure (ERP).

Prerequisites
- Operations lead operator to notify construction manager and safety and loss control leader of the planned offload.
- Suspend all adjacent (within immediate vicinity and downwind) safe work agreements and evacuate the units for the duration of the offload.
- Acid vapor scrubber (03-T-341) is running and ensure pH is > 8.5.
- Corner offload beacon lights must be turned on.

Supporting Documents

P and ID A1-003-B-012, B-12A, B-25.

Steps	Initials	Date and Time

• NOTE: All steps are the responsibility of the UNIT 300 (water treatment plant [WTP]) operator unless otherwise noted.
 • NOTE: Attempt to off load during daylight hours whenever possible.
 • NOTE: The 300 (WTP) operator MUST
     1. WITNESS the hooking up and disconnection of the acid tanker.
     2. REMAIN around BU-300 for the duration of the offload.
     3. ASSUME the role of the first responder in case of an emergency.		

1. PRINT the following: • Critical Safe Work Procedure S15-T 309-OP to fill the acid tanks 03-T309 A/B. • ACID/CAUSTIC OFFLOADING ERP 10-04.	_____	_____
2. ENSURE enough competent personnel are on site to fill all the roles as defined on the ERP. If not, DO NOT proceed until enough personnel are on site.	_____	_____
3. WRITE the names of specific individuals on the ERP.	_____	_____
4. REVIEW with the truck driver: • ERP. • PPE requirements. • Critical safe work procedure S15-T 309-OP to fill the acid tanks 03-T309 A/B.	_____	_____
5. ISSUE a safe work agreement and conduct a hazard assessment with the truck driver to safely unload hydrochloric acid into the acid storage tanks.	_____	_____
6. CHECK waybill to ensure correct product is being delivered and we have sufficient space in our hydrochloric acid storage tanks to accept the delivered volume.	_____	_____
NOTE: Our tanks hold 100 m^3 and are equalized; a load of 28,000 kg will increase our (equalized) tanks by 14–16%.		
7. ISSUE the truck driver a radio on channel 3 and a bump-tested gas monitor.	_____	_____
8. INSTRUCT the truck driver on how to use the radio and gas monitor.	_____	_____
9. ESCORT the driver to the acid tanks and ASSIST the driver in spotting the truck.	_____	_____
10. CONFIRM contents of BU-300 emergency locker. Yellow "pull tight" seal must be in place.	_____	_____
CAUTION: If the yellow pull tight seal is not in place, restock locker prior to moving to step 11. You must advise your supervisor after the offload is complete and the locker is restocked and resealed.		

Supporting Documents

P and ID A1-003-B-012, B-12A, B-25.

Steps	Initials	Date and Time
11. SHOW the truck driver the location of the nearest safety shower(s)/ eye wash station(s).	_____	_____
12. TEST the operation of the safety shower(s)/eye wash station(s); ENSURE the control room received an alarm on the distributed control system (DCS).	_____	_____

Note: This must be the closest shower to the offload point with the easiest access. 3B (in the middle of BU300) **or** 3C (on the east end of BU300).

Steps	Initials	Date and Time
13. INSPECT all fittings and hoses for compatibility and wear.	_____	_____
14. ENSURE driver hooks up the truck discharge hose to the tank fill connection and secures the cam lock fittings with safety wire/pins.	_____	_____
15. ENSURE grounding cable between truck and tank is installed.	_____	_____
16. ENSURE drip pans are located underneath all connections to catch any spills.	_____	_____
17. ENSURE truck driver demonstrates how to operate the emergency shutoff on the truck to 300 (WTP) operator.	_____	_____
18. NOTIFY lead of the emergency rescue team and control room operator that acid offloading is about to commence.	_____	_____
19. OBSERVE the wind direction by using the visual aids (wind socks).	_____	_____
20. ENSURE that the UNIT 300 (WTP) operator is standing upwind prior to commencing offload.	_____	_____
21. OBSERVE the charging of the acid offload lines from the tanker to the tank.	_____	_____
22. TRUCK DRIVER: • Start OFFLOADING the hydrochloric acid. • Must not let the pressure in the trailer exceed the working pressure of the trailer. • Must stay by the product valve while offloading. • When approximately 90% of the product is offloaded close the product valve 50% to slow the product entering the storage tank. • Be prepared to close the valve completely as air starts to pass through.	_____	_____

CAUTION: 300 (WTP) Operator MUST be in offload area during the hookup, the initial offloading, and when disconnecting the tanker truck.

Steps	Initials	Date and Time
23. SHUTDOWN offloading if • HI LEVEL alarm comes in. • Any leaks are observed. • Any unusual condition is observed.	_____	_____

Supporting Documents		
P and ID A1-003-B-012, B-12A, B-25.		

Steps	Initials	Date and Time

CAUTION: Do not allow excessive air to vent into the storage tank. This may cause the product to foam or spray out. Extreme caution is to be used with fiberglass tanks as they will rupture if too much air is discharged through them as the trailer empties. During the last 10%, pinch manual valve to prevent ruptures.

24. TRUCK DRIVER:
- Leave the product valve closed for a few seconds and then quickly open and close it again.
- Use the manual valve to depressurize through the scrubber system. **Ensure this is done slowly to protect the scrubber.**
- When nearing the end of the offload (last 10%), truck driver needs to throttle back and reduce the offload valve to sufficiently prevent overpressuring the tank. This will help with a controlled depressurization.
- BLOW DOWN the truck and lines until no residual pressure is left once the tanker is empty.

NOTE: Truck driver must resist the urge to increase the position of this valve. It will take approximately 35 minutes to depressurize the entire tanker trailer as a controlled blowdown.

CAUTION: If venting/lifting occurs during blowdown, truck driver MUST throttle back the valve position until such time that the vapors disappear.

25. TRUCK DRIVER to
- INFORM 300 (WTP) operator that he/she is ready to disconnect.

26. PROCEED to acid offloading area to witness the hose disconnect.

27. NOTIFY lead of the emergency rescue team and control room operator that acid offloading is complete.

28. INSPECT and RETURN all PPE to appropriate storage locker.

29. OBTAIN the radio and monitor from truck driver.

30. SIGN OFF the safe work agreement.

31. ENSURE all workers in and around BU-300 are notified that ACID offloading has been completed.

32. END OF CRITICAL PRACTICE.

Appendix 3: Sample Code of Practice*

Author:	Approved by:
Revised by:	Signature:
Revised by:	
Date:	
Date:	

Objectives
- To identify and isolate electrically operated equipment for the purpose of undertaking replacement, repair, or maintenance work on the equipment.
- To complement the safety requirements of a safe work permitting process.

Hazards
- Hazardous, combustible atmosphere.
- Incomplete or out-of-date electrical drawings that may not accurately portray all potential sources of electrical energy.
- Inadvertent movement of mechanical equipment.
- Stored electrical energy in the equipment or stored electrical energy in electrical circuits.
- Hazardous layouts: duplicate panels, confined spaces.
- Possible live circuits adjacent to circuits being serviced or maintained or isolated.

Personal Protective Equipment Requirements
- "DO NOT OPERATE" tags with provision for signature, date, and reason for lockout.
- Organization-specific locks, individually keyed and marked, tagged or color coded to identify the organization or work group applying the locks. All locks to be from site supplies.
- Hasps that are capable of receiving a minimum of three locks.
- Lockboxes if multiple locks can be applied. Located on site.
- Chains, bars, clamps, or bonnets that can be secured by a lock for securing valves or other equipment. Located on site.

Supporting Documents and Location
- Safe work agreement.
- Confined space entry code of practice.
- Mechanical zero energy/lockout procedure.
- Federal and provincial regulations.
- Electrical and instrumentation drawings in control room/document room.

Steps	Notes
Working With Open (Dead) Circuits:	
1. Analyze task and each potential energy source.	A qualified licensed electrician or instrument technician is required to identify the potential sources.
2. Identify the point(s) of control to be locked out.	CAUTION: Distributed control system (DCS) can override the programmable logic control (PLC). The point of control is where the energy-blocking, -isolating, or -dissipating devices are controlled.
3. Monitor atmosphere surrounding the equipment where the electrical work is to be performed.	Atmospheric testing must confirm 0% lower explosive limit (LEL) for work activities with the potential for creating an ignition source.
4. Isolate electrical source by shutting off main breaker or individual circuit breaker(s) or disconnecting the circuit from the main or subpanels.	NOTE: All circuits above 480 volts have to be isolated by a qualified electrician. CAUTION: Test the circuit with a voltage meter (set to the proper voltage range for the circuit being tested) to ensure that there is no energy beyond the breaker AND to ensure that the "neutral" circuit is not common to other circuits that are still energized.
5. Apply locks and tags.	NOTE: If disconnect is not designed to contain a padlock, then disconnect the circuit at the main panel. NOTE: Locks must be individually keyed so that only the individual (or equivalent on opposite shift) who installed the lock can remove it. Also, each person of a different trade or function is required to add a lock to the equipment or circuit being serviced or maintained. NOTE: The codes of practice of the organization require a minimum of two locks on each control point. Additional locks for each trade may be locked in a lockbox or added to each control point. Lock in the "off" position.
6. Confirm that the electrical source is isolated at the point where the repair, replacement, or adjustment work is to be performed.	CAUTION: For three-phase circuits, confirm that there is no power on any of the circuits by testing with a voltage meter (set to the proper voltage range for the circuit being tested). NOTE: All circuitry will need to be tested to ensure isolation integrity. • Bump test at the control switch. • Bump test at the control room panel. This will ensure isolation of the correct switching circuit.

Steps	Notes
Working with Live Circuits	
7. Review the steps necessary for performing hot work in a hazardous atmosphere.	Refer to the safe work agreement process of the organization for hot work. Attach appropriate tags and lockouts to inform others of the work in progress.
8. Analyze the situation to understand the potential changes in the circuits or instruments being monitored or calibrated online.	CAUTION: Isolate any remote control capability during the monitoring or calibration. Minimize the influences on the circuits or instruments being monitored to stabilize the conditions at the site.
9. Conduct the maintenance or servicing work to be performed.	CAUTION: Monitor energy sources during work to ensure that isolation integrity is maintained. CAUTION: Do not leave live circuits in an unsafe condition if work carries over a break (i.e., replace covers, panel doors, lockout, etc.). NOTE: If work carries over a break period that results in the workers leaving the site OR it is extended to another shift or calendar day(s), then it is essential to repeat step 7 after each interruption in the work activities.
10. Inspect the integrity of the electrical equipment following the completion of maintenance or servicing work.	Ensure that all connections have been replaced, bolts are tightened, instrument covers or parts are properly installed.
11. Remove the lockouts of the control points.	Prepare the equipment for recommissioning if electrical circuits were deenergized.
12. Monitor for normal operation.	

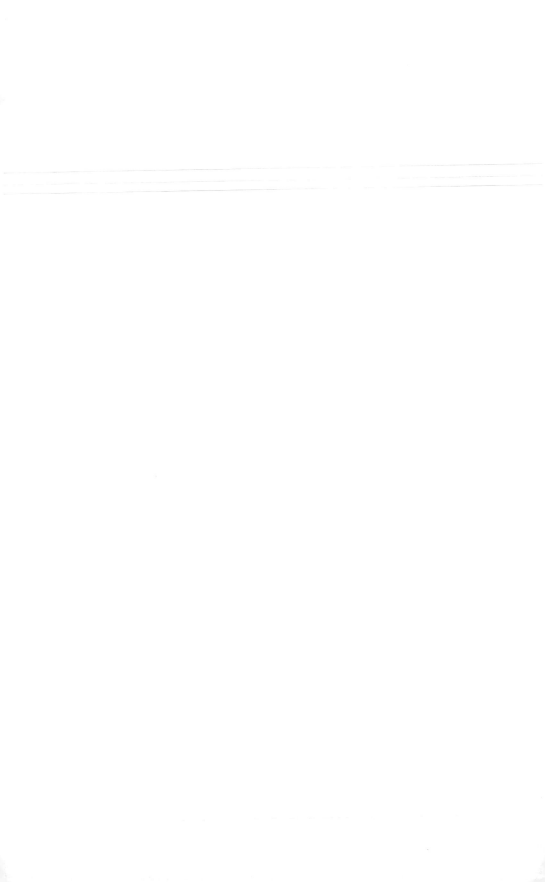

Appendix 4: System Turnover Responsibilities and Method*

1.0 Purpose and Scope

    1.1 To define the interfaces between construction and commissioning organizations and commissioning and operations organizations and the responsibilities of all groups involved in the project.

    1.2 To provide a method to ensure that all equipment or systems have been tested and confirmed operational consistent with design specifications before the system is turned over to the operations organization for final ownership.

    1.3 To provide a verifiable audit trail via documentation that confirms all systems (and equipment) are ready for safe and continuous operation.

2.0 Definitions

    2.1 Construction inspections and checks

        Inspections and checks performed by the construction organization to confirm that required construction activities are completed satisfactorily and consistent with design prior to turnover to the commissioning organization for preoperational testing and initial operations.

    2.2 Loop check

        A verification of the integrity of control circuits by simulation or manipulation of every contact or device within the control circuit with control power applied.

    2.3 Instrument calibration

        Adjustment of instruments and control devices utilizing predetermined values and verification of an acceptable quantitative level accuracy. This includes the verification of the functional integrity of devices or instruments that are not adjustable by normal means.

2.4 Functional loop check

A set of tests is performed for the purpose of verifying the functional integrity of an instrumentation control loop and verifying that instruments will meet design criteria for systems (static head, reset capabilities, etc.).

These loop checks are performed with all loop components installed, electronic loop energized, or pneumatic loop pressurized. This includes a loop check from the primary device to the distributed control system (DCS) control board and from the DCS back to the final device.

2.5 Final acceptance

Satisfactory completion of the performance test and the facility confirmed operational to the satisfaction of the operations organization.

3.0 Participating Groups

3.1 Construction organization

The group charged with the responsibility for constructing the facility consistent with design specifications.

3.2 Commissioning organization

The group charged with the responsibility for testing equipment and machinery to ensure proper construction and equipment functionality within systems consistent with design requirements before turning over equipment or systems to the operations organization for steady-state operations.

3.3 Operations organization (owner)

Consists of the operations leader and delegates qualified to accept turned-over systems.

4.0 Responsibilities

4.1 Construction organization

4.1.1 This organization provides scheduled completion and delivery to the commissioning organization, all systems consistent with milestone deliverables.

4.1.2 The organization participates in the system turnover process.

4.1.3 The organization oversees inspection activities performed in accordance with the construction control and completion program for quality assurance/quality control (QA/QC) verification.

4.1.4 The organization ensures systems are installed and mechanically completed to the extent required for each milestone activity or as requested by the commissioning organization before turnover for commissioning and preoperational testing. Any equipment or component found to be installed improperly or defective may be turned back to the construction organization for remedial actions.

4.1.5 The construction leader shall be the interface between the commissioning organization and the construction organization. He or she shall ensure schedule completion of construction activities affecting preoperational testing and system turnover to the operations organization.

4.1.6 The construction organization coordinates system deficiency resolution after system turnover to the commissioning organization and resolves all construction completion items in a timely manner.

4.1.7 The construction organization ensures all equipment until turned over; meets vendor requirements consistent with instruction manuals. Such activities are to be archived and turned over to the commissioning organization.

4.2 Commissioning organization

4.2.1 Commissioning leader

This leader ensures that commissioning engineering procedures required for the testing and commissioning of equipment and systems are prepared and reviewed prior to equipment startup.

The commissioning leader shall coordinate all commissioning activities to protect the safety of personnel and to prevent damage to equipment and the environment. The commissioning leader is responsible for the overall direction and conduct of testing on plant equipment and systems. He or she assumes primary responsibility for scheduling and directing the efforts of those assigned to him or her in the performance of testing and commissioning activities. He or she shall coordinate the interface activities of the commissioning, construction, and operations organizations and support services to fully commission each system of the facility.

4.3   The operations organization is responsible for the following activities:

4.3.1   Assists in the operation of all permanent plant equipment to support the commissioning activities. The operations organization, where possible, will integrate into the commissioning organization to support commissioning activities. These personnel will be returned to the operations group once the facility assumes *hot* status with the introduction of process hazards for continuous operation, such as temperatures, pressures, and fluids into the facility.

4.3.2   Participates in the system turnover procedure as outlined in this document and receives systems that have been proved and are ready for turnover to operations.

5.0   Inspections and Testing Responsibilities

5.1   Construction completion phase

During the construction of the individual systems, construction supervisors and engineers shall perform necessary and required inspection to ensure that completed installations are in accordance with the engineering and design information and specifications.

5.1.1   Electrical

All electrical wiring (power, control, and instrumentation) shall be installed and terminated in accordance with design requirements. The construction leader shall ensure that all work is done using competent and qualified personnel and work performed is of an acceptable standard. All testing shall be performed in accordance with a relevant procedure and shall be appropriately documented.

5.1.2   Mechanical

All mechanical equipment shall be installed in accordance with design standards. Final coupling alignment shall be done after commissioning engineers have bumped tested and run the associated motor satisfactorily. All equipment lubrication shall be done during this construction phase before turnover. Lubrication and alignment check shall be documented for documentation turnover. All rotating equipment operational problems associated with coupling misalignment or pipe stress will be the responsibility of the construction group, and associated equipment will be returned to the construction organization for corrective action.

5.1.3 Piping

All piping shall be installed in accordance with design standards. Before turnover to the commissioning organization, the construction organization shall conduct piping system hydrostatic tests that are required. The construction organization shall complete all piping changes required to obtain correct coupling alignment.

5.1.4 Instrumentation

All instrumentation devices and piping shall be installed in accordance with design standards. All tubing hydrostatic testing shall be conducted by the construction organization before system turnover.

5.2 Preoperational testing phase

The commissioning leader and his or her team shall direct and supervise all system verification checks and testing personnel during this phase.

5.2.1 Electrical testing

Electrical testing and verification scope shall include all high-voltage switchgear, transformers, protective relays, 120-volt AC (alternating current) control circuits, breaker control circuits, batteries, and DC (direct current) control circuits. All electrical functional circuit verification and testing shall be conducted in accordance with appropriate electrical test procedures.

Verification and testing methods and practices will include highlighting (with a colored marker) all electrical drawings to confirm verification is completed and filling out appropriate electrical test procedures data sheets for each test conducted.

5.2.2 Instrumentation testing

Instrument calibration: Instrument testing and verification scope shall include calibration. After each instrument is calibrated, a calibration data sheet shall be retained as verification of the calibration and for inclusion in the turnover documentation package.

DCS control circuits: DCS instrument operations (IO) circuitry shall be tested in accordance with the following:

All logic action will be verified and documented. Any logic previously verified during the factory acceptance test need not be reverified at site on agreement with construction leader and commissioning leader. Documentation is required. Any logic introduced or changed at site must be verified by testing and appropriately documented.

5.2.3 Systems testing

The system commissioning organization is accountable for performing and directing all system commissioning activities. Scope of work will also include system inspections, chemical cleaning, and steam blow activities if identified in the schedule. The initial start-up and operation of all systems are the responsibility of the commissioning organization.

5.2.4 System completion

The commissioning leader shall maintain a system completion checklist to track and confirm when a system is complete. This checklist will also be included in each system turnover document package binder.

5.3 System start-up/operational phase

The final system start-up and commissioning after testing shall be accomplished by the commissioning organization with the assistance of the operations organization. This phase includes a validation of the outcome of all tests performed during commissioning to ensure satisfactory continuous and commercial operation. This phase shall include (a) placing all systems into continuous operation and (b) identifying and resolving any missed continuous operational problems. Until final system turnover, the commissioning organization retains ownership of the system.

6.0 System Identification

All system correspondence, technical data, and information shall be filed using the system boundary identification list and index. All systems turned over from the construction organization to the commissioning organization shall be identified by a system turnover number. A similar process is required for turnover from the commissioning organization to the operations organization.

7.0   System Turnover Method

    7.1   Turnover sequence summary

        7.1.1   Turnover from the construction organization to the commissioning organization occurs on mechanical completion and system readiness for testing.

        7.1.2   When the commissioning leader is satisfied that the system described is ready for preoperational testing, the commissioning leader shall sign the system turnover form (Figure 7.2), formally accepting the system for preoperational testing.

        7.1.3   The commissioning organization shall conduct a thorough inspection of the system, perform the required commissioning testing, and perform start-up of the system.

        7.1.4   The commissioning organization shall determine that a system is ready for routine operation, prepare a complete system turnover documentation package, and sign the release to operations section of the system turnover form (Figure 7.2), releasing the system to the operations organization for ownership.

        7.1.5   The commissioning organization updates the punch list, signs the system turnover form (Figure 7.2), and submits the turnover package to the operations organization.

        7.1.6   The operations organization inspects the system, reviews the system turnover package, and confirms acceptance of the system by signing the system turnover form.

    7.2   Detailed system turnover process

        The following method shall be used to turn over a completed system from construction to commissioning and then from commissioning to owner.

        7.2.1   Commissioning organization initial system walkdown

            At least 1 week prior to hydrotesting a system, the construction leader shall notify the commissioning leader that the system is approaching mechanical completion and is ready for a preliminary walkdown to determine system completeness and operability (initial punchlisting).

        7.2.2   System punchlisting

            Representatives of the construction organization, commissioning organization, and operations organization perform a preliminary construction system walkdown. The goal is to identify all construction deficiencies that would have an impact on system testing and that require remedial actions before turnover of the system to the operations organization. The construction organization shall maintain the list of identified deficiencies and provide copies to all participants of the system walkdown. The three stakeholders will perform a final construction walkdown after the items from the preliminary walkdown are resolved to completeness. This walkdown will also identify all other system deficiencies to be included in the system punch list. The construction leader or commissioning leader may initiate the construction turnover form. The construction leader shall ensure the latest copies of the redlined drawings are turned over to the commissioning organization at the time of the system turnover. Prior to turnover of a system to the construction organization, the construction leader shall coordinate resolution of all A deficiency construction punch list items. The completed turnover form, including the system punch list, shall be forwarded to the commissioning leader.

7.2.3 The commissioning organization shall perform an inspection of the system, verifying A deficiency resolution prior to accepting the system. When the commissioning organization is satisfied that the system described on the system turnover form is ready for preoperational testing, he or she shall sign the turnover form, accepting the system for preoperational testing. The commissioning organization shall issue appropriate communication and warning notices regarding commissioning-related hazards during testing.

7.2.4 The commissioning organization conducts a thorough inspection of the system and performs the required commissioning testing and start-up of the system. During the course of testing, information detailing existing punchlist items completed or new punch list items identified will be updated on the punch list. The commissioning organization shall determine that a system is ready for routine operation, prepare a complete system turnover documentation package, and sign the release to operations portion of the system turnover form for final turnover to the operations organization.

7.2.5   The Construction Leader reviews the turnover package and updates and approves the latest punch list.

7.2.6   The operations organization shall perform a review of the system documentation. This organization will also perform a system walkdown jointly with the commissioning organization to provide a final update to the punch list prior to final acceptance of the system. When the operations organization is satisfied that the system or subsystem described on the system turnover form is satisfactory with no A deficiencies, the system turnover form is signed as an acknowledgment that the system is acceptable for routine operation.

7.2.7   System Punch List

An up-to-date system punch list must accompany each system turnover or partial turnover. The punch list for a particular system will be stored in the respective system turnover package binder. The A deficiencies must be completed before any system turnover from one organization to another.

The B deficiencies must be completed before final acceptance by the owner's representative. Alternatively, at the discretion of the operations organization, certain B deficiencies may be resolved after turnover of the system.

Appendix 5: Checks and Tests Performed by the Operations and Commissioning Organizations*

Discipline Required	Operator	Electrical	Instrument	Mechanical	Engineering	Loss Management	Comments
Piping							
Design size, spec, and support					X		
High- and low-point drains and vents	X						
Flange bolts installed correctly				X			
Gasket installed properly				X			
Design gasket material/ insulation				X			
Spec blinds installed, orientation correct	X						
Line adequately heat traced	X						
Line adequately insulated	X						
Adequate room for expansion of pipe, flanges, and take-offs (no stress/strain)	X				X		
Access to instruments: Valves/ transmitters, safety systems such as fire eyes			X				
Adequate clearance between pipes, for low-point drains/ purges	X						
Line layout, trip/overhead hazards	X						
Line gradient adequate: per design					X		
Welding quality: workmanship					X		
Pipe hangers properly installed				X			
Reducer orientation					X		
Restrictive and flow orifices properly installed			X				

---

* Copyright Suncor Energy Inc.; approval from Suncor Energy Inc. is required to reproduce these data.

185

Discipline Required	Operator	Electrical	Instrument	Mechanical	Engineering	Loss Management	Comments
One-way check valve orientation	X						
Valve orientation for maneuverability and safety	X						
Control valve orientation			X				
Sample system installed to design standards					X		
Grounding installed on instrumentation			X				
Grounding system for plastic pipe	X						
Strainers installed				X			
Workplace hazardous materials information system (WHMIS) identification						X	

Pressure Safety Valves (PSVs), Manual and Automatic Valves

Accessible for safe operation	X						
All bonnets and flanges installed			X				
Access to bonnets and flanges for maintenance			X				
Packing installed and accessible			X				
Designed gaskets used				X			
Hand wheel or valve handle installed and secure	X						
Valve body heat traced and insulated	X						
Lubricator oil level correct			X				
Electrical conduit sealed		X					
Limit switch indicators visible and accessible			X				
Plugs installed on all drains/ bleeds	X						
Air supply piped accurately with regulators, filters, and screens on exhaust ports			X				
Exhaust ports directed to safe area			X				
Correct torque setting on electric valves actuators		X					
Valve positioned appropriately	X						

Discipline Required	Operator	Electrical	Instrument	Mechanical	Engineering	Loss Management	Comments
PSV setting matched against P&ID (piping and instrumentation drawings) (design)	X						
Accessible for maintenance				X			
Rupture disk installed if required			X				
PSV discharge piped safely and to proper system (high pressure (HP)/low pressure (LP) flare)	X						
Vent plugs in or out as required by PSV type	X						
Discharge piping adequately supported (not supported by bolting to PSV)	X						
Certification date on PSV current	X						
Heat Exchangers, Vessel, and Towers							
Grouting completed properly				X			
Anchor bolts installed and secured				X			
Vessel grounded as per design				X			
Vessels insulated/heat traced	X						
Bridles and instrumentation insulated	X						
Flanges done up properly				X			
Internal checks completed				X			
Manway access properly secured and accessible				X			
Vessel anchored to ensure proper directional growth				X			
Name plates installed on all vessels	X						
Key internal levels marked externally (e.g., weirs/skims)					X		
Thermal relief systems installed	X						
Safety guards installed on all rotating equipment (i.e., fin fan coolers)				X			
Adequate access for removal and cleaning of exchanger bundles: maintainability				X			

Discipline Required	Operator	Electrical	Instrument	Mechanical	Engineering	Loss Management	Comments
Rotating Equipment							
Coupling installed properly				X			
Safety guards installed on all rotating equipment						X	
Suction installed and orientated correctly				X			
Grouting completed properly				X			
Anchor bolts installed and secured				X			
Drain troughs installed	X						
Equipment grounded as per design		X					
Flanges bolts/gaskets done up tightly				X			
Accessibility for maintenance				X			
PSV on positive displacement pumps: settings and certification dates verified	X						
Cold and hot alignments completed				X			
Direction of rotation checked and confirmed right		X		X			
Sealing systems installed where applicable				X			
Preservation fluids replaced with design lubricants for continuous operations				X			
Lubrication levels confirmed correct				X			
Instrumentation							
Orifice plates orientated and installed correctly			X				
Tapping points for instruments correct: impulse lines, top; gas flow, side; liquid flows, bottom			X				
Transmitters anchored securely (above for gas with slope and below for liquid)			X				
Level gauge locations relative to instrumentation adequate	X		X				

Discipline Required	Operator	Electrical	Instrument	Mechanical	Engineering	Loss Management	Comments
Sufficient range local gauges installed	X		X				
Control valves in vapor service allow for condensate drainage	X		X				
Purge lines available on fire eyes			X				
Regulators stamped to correct settings			X				
Analyzers set up for adequate flow and venting			X				
Instrument and cabinets for area classification rating verified (class/division)		X	X				
Seals poured properly cable seals							
Non-information systems (ISs) and IS cables in separate junction boxes (JBs)		X	X				
JBs checked and verified suitable for hazards classifications		X				X	
Local indicators (lights/horns) mounted in proper area						X	
Electrical							
Power transformer checks completed		X					
Emergency lighting adequate		X					
Adequate and accurate identification on JBs/switch gears and equipment		X					
Hazard identification on panels: adequate for warning against electrical shocks and electrocution		X				X	
Wiring is confirmed against drawings and as-builts created		X					
Cables gland terminated for classification and environment		X					
Ground straps installed where required		X					
Grounding of motor and switch gears confirmed adequate		X					

Discipline Required	Operator	Electrical	Instrument	Mechanical	Engineering	Loss Management	Comments
General							
Egress routes established						X	
Lighting in buildings and outdoors adequate		X				X	
Emergency lighting adequate						X	
Signage adequate to warn workers of hazards						X	
Hazard identification correctly identified						X	
Noise-level testing consistent with regulatory requirements						X	
Protective system layout plan and equipment in place and available for use						X	

Appendix 6: System Turnover Documentation and Control*

System Turnover Documentation and Control

The *system turnover documentation* (STD) shall consist of the following relevant documents defined in this list and must be presented to the operations organization at the final turnover of each system. The STD must be made available at least 5 working days prior to the system turnover for review by the operations organization. Both the construction and the commissioning organizations will be responsible for compiling and presenting the STD package to the operations organization.

The STD package must contain the following sections for each system turned over.

The STD package shall consist of the following:

Section

1.0 Section 1: Documentation

    1.1.1 Updated and Final System Description

    1.1.2 Associated Equipment List

    1.1.3 System Turnover Marked Up P&IDs (Piping and Instrumentation Drawings)

    1.1.4 System Turnover Single Line Diagrams

    1.1.5 System Turnover Instrument Index

    1.1.6 Shutdown Keys

    1.1.7 Commissioning to Operations Checklists

2.0 Section 2: Mechanical

    2.1.1 Construction Check Sheets

    2.1.2 Commissioning Check Sheets

---

* Copyright Suncor Energy Inc.; approval from Suncor Energy Inc. is required to reproduce these data.

3.0   Section 3: Piping
    3.1.1   Construction Check Sheets
    3.1.2   Commissioning Check Sheets

4.0   Section 4: Electrical
    4.1.1   Construction Check Sheets
    4.1.2   Commissioning Check Sheets

5.0   Section 5: Instrument
    5.1.1   Construction Check Sheets
    5.1.2   Commissioning Check Sheets

6.0   Section 6: Commissioning Procedure

7.0   Section 7: Equipment Maintenance Records
    7.1.1   Lube Oil Commissioning Records
    7.1.2   Equipment Rotation Records
    7.1.3   Other Records

8.0   Section 8: Vendor Information

9.0   Section 9: Other Vendor and Related Correspondence

10.0   Punch Lists

11.0   Control Software

12.0   Outstanding Engineering Queries

13.0   Loss Management Checklists

14.0   Management of Change (MOC) Approvals with Supporting Rationale and Risk Review and Mitigation Actions

15.0   Location and Access of Critical Operating Spares on a System Basis

Document Control

A master copy of P&IDs, single line diagrams, line designation tables (LDTs), and shutdown keys will be maintained on site by the construction and commissioning organizations. Changes to the P&IDs will be subject to an MOC process. Redlined marks will be made on these master drawings during the construction and commissioning stages. Each system will be turned over to the operations organization with a complete set of 11 × 17 P&IDs with the attached punch lists specific to the system. A copy of the master shutdown keys, single lines, and LDTs specific to the system will also be turned over to the operations organization with each system turnover. All master drawings will remain with the construction and commissioning organizations until all systems are turned over to the operations organization. At this stage, a copy of all master drawings will be presented to the operations organization for their use and control. Redlined marked-up drawings will be reissued to drafting, and a final as-built set of the master documents will be issued to operations within an agreed time frame.

Glossary of Terms

Critical practices: Work practices that, if not followed as defined in the practice, can result in severe injuries or death to workers or damage to the environment or plant property and equipment. A critical practice is very much like an SOP and must be followed precisely as defined in the practice. Critical practices like SOPs must also be updated at an approved frequency.

Criteria: Those elements that contribute to the readiness of any of people, process, and system component readiness for any particular milestone are its criteria. For example, the criteria for people readiness for a particular milestone may include training completion, availability of SOPs, COPs, and critical practices, among other specific people-related requirements.

Design basis memorandum (DBM): A DBM is a document that defines the scope of a project and the technology to be adopted by the project. The DBM will define the basis on which the engineering design is developed. Types of equipment, mode of application, redundancy, operating pressures and temperatures, types of materials, and process resident time will all be developed during this stage.

Deficiencies: Deficiencies represent incomplete work on a system that may prevent a system turnover to either the commissioning or operations organization. Deficiencies can be classified into A and B deficiencies. The A deficiencies refer to incomplete work that can have an impact on the safety of personnel and may result in damage to plant, property and equipment, or the environment. In addition, A deficiencies have an impact on the ability of a project to deliver on planned outputs. The A deficiencies must be addressed before a system can be turned over from one organization to another. The B deficiencies refer to incomplete work that may have an impact on the safety of personnel or the output of a project. However, B deficiencies can be mitigated against using administrative, procedural, and other controls to mitigate risks and hazards identified until an adequate or permanent remedy is found. The B deficiencies may include incomplete work, such as insulation, labeling, heat tracing, and any other condition that may be remedied through supplemental means.

Debottlenecking: This is a study designed to identify limitations across a facility or project that prevent the project from achieving its planned output levels. A debottlenecking process is often undertaken to augment the performance of the project to meet disclosed design standards. The study is followed up with appropriate remedial actions (including engineering redesign) to enable the facility to produce at planned levels.

Front-end engineering design (FEED): FEED refers to the engineering and design work completed as part of preproject planning done after conceptual business planning and prior to detailed design.

Frontline leaders and supervisors: These individuals are leadership personnel who generally interface with workers at the first level of the organization in a leadership or supervisory role. The term can be broadened to include all leaders and supervisors who are required to influence the behaviors of workers at the front line to get work done. This group of leaders and supervisors is influential in the way (safely, timely) physical work is done at the front line and the amount of work (welding, cleaning, installing, etc.) that is completed on a daily basis. Frontline leaders and supervisors may extend two to three levels up from the frontline workers.

Go-no-go decision: This is an objective decision made by project leaders to proceed from one milestone to another. This decision is dependent on the overall milestone readiness defined by its components readiness. A decision may be made to progress from one milestone to another in the absence of full readiness of the previous milestone. However, project leaders must preferentially allocate resources to bring the readiness of the previous milestone to the acceptable level of readiness for supporting the activities of the new milestone.

Greenfield operations: Greenfield projects or operations refers to brand new facilities that are generally free from constraints resulting from prior operations at a site.

Lean workforce: A lean workforce refers to one in which personnel are required to multitask and are often multiskilled. Lean workforces depend on automated systems and self-regulation. Lean workforces promote mutually agreed rules across departments and will depend on full access to information, tools, and spare parts to maintain continuous operations during low work periods. For example, a warehouse may be unattended during night shifts or over the weekend, during which periods operating personnel may be required to access the warehouse for replacement parts or consumable items according to mutually agreed rules.

Management of change (MOC): This is a procedure required for managing risks associated with changes made in an industrial process environment. MOC procedures require the right levels of expertise to evaluate the impact of a proposed change on an operation before a change is approved. The procedure requires the right level of authority for approval of the proposed change and a mechanism for ensuring newly introduced risks are properly assigned and mitigated once the proposed change is made.

Mechanical completion (completeness): This term refers to the installation of all vessels, equipment, and machinery and all mechanical interconnections (piping) of a system. The system is physically available, but testing for integrity has not yet been completed on the system.

Milestones: These are a series of significant actions or events during project execution that signal a change in the way work is conducted at the work site or a major achievement. Milestones are defined early in the project execution stage and are used to gauge and demonstrate tangible progress toward completion of the project.

People: People are all personnel involved in the project execution stage of the project. Included are personnel who make up the construction, commissioning, and operations organizations as well as other personnel who may support the project execution.

PP&S: This acronym refers to people, processes, and systems in the readiness process. *See* People, Process, and System entries.

Process: This includes processes required for supporting project execution activities and the steady-state operations requirements of the project.

Project leaders: These individuals are the Project Manager and supporting senior leaders responsible for the construction, commissioning, and operations organizations during project execution.

Punchlisting: This process identifies deficiencies in a system being turned over from the construction to the commissioning organization or from the commissioning organization to the operations organization. Punchlisted items can be classified into A deficiencies or B deficiencies.

Readiness: This is an objective method used for determining the completion of people, process, and system activities during the execution stage of the project cycle. Readiness refers to the completeness of activities that enables project leaders to proceed from one milestone to another.

Retrofitting: Retrofitting relates to updating an asset or facility after the execution stage of a project is completed and the project has begun steady-state operations. It refers to the process of installing equipment or machinery that may have been left out earlier for one reason or another and may now be deemed essential for the operation.

Scope (project): The scope of a project refers to all work to be completed as defined in the DBM of the project. Sharma and Lutchman (2006) suggested the scope of a project refers to the "equipment and machinery to be provided and the work to be done and is documented by the contract parameters for the services to be provided by a service provider, or contractor" (p. EST. 16.1).

Showstoppers: In the readiness process, failure to achieve an acceptable level of readiness can result in a failure to progress from one milestone to another. This condition is known as a showstopper. An example of a showstopper is inadequate people readiness (personnel not trained or critical practices for managing power at the site not completed) and the movement from milestone 1 (M1) to milestone 2 (M2), ready to introduce permanent power to the site.

Situational leadership: This model for leadership requires leaders to individually consider the work maturity level of the worker such that appropriate leadership styles can be applied for the specific maturity level of the worker. Situational leadership requires leaders to accurately identify the maturity level of the worker and to use leadership behaviors consistent with the maturity level of the worker. The situational leadership model was developed by Paul Hershey and Ken Blanchard (Bass, 1990) and has been one of the more successful models for developing the workforce.

Standard operating procedures (SOPs): These procedures are developed for guiding inexperienced workers with complex work activities in a process environment or facility. SOPs also guide experienced workers on unfamiliar jobs. SOPs are generally derived from the operating manuals provided by vendors or may be developed based on experiences of operating personnel. SOPs must be updated at a facility-approved frequency to ensure improvements are incorporated into the procedure over time.

Systems: The mechanical equipment, machinery, interconnecting pipes, valves, hardware, and accessories for any particular product or input at the facility are its systems. For example, the instrument air systems may include the hardware and infrastructure required for the supply and distribution of dry compressed air to all control valves at the facility.

System walkdown: This is the physical process of examining a system to be turned over from one organization to another to identify the completeness of the system before turnover can occur. The process involves using the piping and instrumentation diagrams to ensure all components of the engineering design are included in a physical system of piping, equipment, and machinery.

Training matrix: This simple matrix represents the training required by personnel on one axis and the personnel to be trained on the other. The intersection point between specific training course and an individual may be color coded to show visibly whether an individual does not require training in that area or is untrained, trained, or competent.

Turnover process: The turnover process is a legal process among organizations involved in project execution, and it involves the changing of ownership of systems from one organization to another until final ownership is passed to the operations organization. The turnover process is accompanied by all documentation (generally two hard copies in binders) with all relevant testing results and checks for integrity performed on the system.

References

Agrusa, J., & Lema, J. D. (2007). An examination of Mississippi gulf coast casino management styles with implications for employee turnover. *UNLV Gaming Research and Review Journal, 2007, 11*(1), 13–26. Retrieved April 11, 2007, from EBSCOHost database.

Alberta Economic Development Authority. (2004, December 10). Mega project excellence: Preparing for Alberta's legacy. Retrieved August 27, 2007, from http://www.alberta-canada.com/aeda/pdf/MegaProjectExcel_Dec102004.pdf.

Allen, W. R., Drevs, R. A., & Rube, J. A. (1999). Reasons why college-educated women change employment. *Journal of Business and Psychology, 14*(1), 77–93. Retrieved April 17, 2007, from EBSCOHost database.

Ashworth, M. J. (2006). Preserving knowledge legacies: Workforce aging, turnover and human resource issues in the U.S. electric power industry. *International Journal of Human Resource Management, 17*(9), 1659–1688. Retrieved November 24, 2007, from EBSCOHost database.

Austin, I. (2008, May 1). Canadians investigate death of ducks at oil-sands project. *New York Times*. Retrieved January 22, 2009, from http://www.nytimes.com/2008/05/01/business/worldbusiness/01sands.html.

Bass, B. M. (1990). *Handbook of leadership theory, research, and managerial applications* (3rd ed.). New York: Free Press.

Blanchard, K. (2004). Leadership and the bottom line. *Executive Excellence, 21*(9), 18. Retrieved July 26, 2008, from EBSCOHost database.

Blanchard, K. (2008). Situational leadership. Adapt your style to their development level. *Leadership Excellence, 25*(5), 19. Retrieved July 25, 2008, from EBSCOHost database.

Branch, S. (1998). You hired 'em. But can you keep 'em? *Fortune, 138*(9), 247–250. Retrieved August 27, 2007, from EBSCOHost database.

Carter, R. A. (2007). Canadian mining: A crowded house. *Engineering and Mining Journal, 208*(7), 68–77. Retrieved November 10, 2007, from EBSCOHost database.

Conklin, M. H., & Desselle, S. P. (2007). Job turnover intentions among pharmacy faculty. *American Journal of Pharmaceutical Education, 71*(4), 1–9. Retrieved November 17, 2007, from EBSCOHost database.

Cooper-Hakim, A., & Viswesvaran, C. (2005). The construct of work commitment: Testing an integrative framework. *Psychological Bulletin, 131*, 241–259. Retrieved October 28, 2006, from EBSCOHost database.

Daly, C. J., & Dee, J. R. (2006). Greener pastures: Faculty turnover intent in urban public universities. *Journal of Higher Education, 77*(5), 776–803. Retrieved April 22, 2007, from EBSCOHost database.

Davis, J. H., Schoorman, F. D., Mayer, R. C., & Hwee Hoon, T. (2000). The trusted general manager and business unit performance: Empirical evidence of a competitive advantage. *Strategic Management Journal, 21*(5), 563–576. Retrieved November 6, 2006, from EBSCOHost database.

Desselle, S. P. (2005). Job turnover intentions among certified pharmacy technicians. *Journal of the American Pharmacists Association, 45*(6), 676–683. Retrieved May 28, 2007, from http://www.medscape.com/viewarticle/518371.

Energy Resources Conservation Board (ERCB) Investigation Report. (2008). MEG energy corp. steam pipeline failure, May 25, 2007. Retrieved March 1, 2009, from http://www.ercb.ca/docs/documents/reports/IR_20080902_MEG_Energy.pdf.

Facione, P. (2006). Critical thinking: What it is and why it counts. Retrieved January 19, 2009, from http://nsu.edu/iea/image/critical_thinking.pdf.

Fields, D., Dingman, M. E., Roman, P. M., & Blum, T. C. (2005). Exploring predictors of alternative job changes. *Journal of Occupational and Organizational Psychology, 78*(1), 63–82. Retrieved October 1, 2006, from EBSCOHost database.

Finegold, D., Mohrman, S., & Spreitzer, G. M. (2002). Age effects on the predictors of technical workers' commitment and willingness to turnover. *Journal of Organizational Behavior, 23*(5), 655–675. Retrieved October 21, 2006, from EBSCOHost database.

Gentry, W. A. (2006). Human resource officers' opinions of their own companies and "the big three": A qualitative study profiling businesses in a southeastern city. *Organization Development Journal, 24*(2), 33–42. Retrieved April 26, 2007, from EBSCOHost database.

Green, R. (2004). Hiring so you won't be firing. *Journal of Organizational Excellence, 23*(3), 99–100. Retrieved October 15, 2006, from EBSCOHost database.

Guthrie, J. P. (2001). High-involvement work practices, turnover, and productivity: Evidence from New Zealand. *Academy of Management Journal, 44*(1), 180–190. Retrieved November 9, 2007, from EBSCOHost database.

Haggett, S. (2008, April 30). Syncrude probed as hundreds of ducks die in oilsands. *Financial Post.* Retrieved January 23, 2009, from http://www.financialpost.com/story.html?id=482872.

Halama, L. M., Kelly, D., Rideout, S., & Robinson, P. (2004, May 20). Leading indicators best practices presentation. Construction Owners Association of Alberta. Available March 21 from http://www.coaa.ab.ca/LinkClick.aspx?link=pdfs/Leading%20Indicators/COAA%20Leading%20Indicators%20-%20Best%20Practices%20Conference%20Presentation.pdf&tabid=154.

Harris, K. J., Kacmar, K. M., & Witt, L. A. (2005). An examination of the curvilinear relationship between leader–member exchange and intent to turnover. *Journal of Organizational Behavior, 26*(4), 363–378. Retrieved April 11, 2007, from EBSCOHost database.

Harris, R. (1998, July 1). Introduction to creative thinking. Retrieved January 20, 2009, from http://www.virtualsalt.com/crebook1.htm.

Hemdi, M. A., & Nasurdin, A. M. (2006). Predicting turnover intentions of hotel employees: The influence of employee development human resource management practices and trust in organization. *Gadjah Mada International Journal of Business, 8*(1), 42–64. Retrieved November 9, 2007, from EBSCOHost database.

Hoffman, M. (2007). Employee retention begins with effective leadership. *Hotel and Motel Management, 222*(1), 58–58. Retrieved April 16, 2007, from EBSCOHost database.

Hopkins, S. M., & Weathington, B. L. (2006). The relationships between justice perceptions, trust and employee attitudes in a downsized organization. *Journal of Psychology, 140*(5), 477–498. Retrieved November 24, 2007, from EBSCOHost database.

Institute of Management and Administration. (2007). Who helps HR executives plan compensation? *Report on Salary Surveys, 1*(7), 1–15. Retrieved April 8, 2007, from EBSCOHost database.

Korkmaz, M. (2007). The effects of leadership styles on organizational health. *Educational Research Quarterly, 30*(3), 22–54. Retrieved April 16, 2007, from EBSCOHost database.

Lambert, E. G., Hogan, N. L., & Barton, S. M. (2001). The impact of job satisfaction on turnover intent: A test of a structural measurement model using a national sample of workers. *Social Science Journal, 38*(2), 233–250. Retrieved October 28, 2006, from EBSCOHost database.

Little, P. L. (2006). The high cost of under-skilled labor. *Industrial Maintenance and Plant Operation, 67,* 10. Retrieved October 24, 2006, from EBSCOHost database.

Lockwood, N. R. (2007). Leveraging employee engagement for competitive advantage: HR's strategic role. *HR Magazine, 52*(3), 1–11. Retrieved April 26, 2007, from EBSCOHost database.

Lutchman, C. (2008). Leadership impact on turnover among power engineers in the oil sands of Alberta. Dissertation, University of Phoenix.

Madlock, P. E. (2008). The link between leadership style, communicator competence, and employee satisfaction. *Journal of Business Communication, 45*(1), 61–78. Retrieved July 25, 2008, from EBSCOHost database.

Mansell, A., Brough, P., & Cole, K. (2006). Stable predictors of job satisfaction, psychological strain, and employee retention: An evaluation of organizational change within the New Zealand customs service. *International Journal of Stress Management, 13*(1), 84–107. Retrieved April 22, 2007, from EBSCOHost database.

Meisinger, S. (2006). Talent management in a knowledge-based economy. *HR Magazine, 51*(1), 10. Retrieved April 8, 2007, from EBSCOHost database.

Michalko, M. (2000). Four steps toward creative thinking. *The Futurist, 4*(3), 8–21. Retrieved January 20, 2009, from EBSCOHost database.

Owens, P. L. (2006). One more reason not to cut your training budget: The relationship between training and organizational outcomes. *Public Personnel Management, 35*(2), 163–172. Retrieved April 12, 2007, from EBSCOHost database.

Phani, T. A. (2006). Causes and consequences of high turnover by sales professionals. *Journal of American Academy of Business, Cambridge, 10*(1), 137–141. Retrieved November 12, 2007, from EBSCOHost database.

Pisarski, A., Brook, C., Bohle, P., Gallois, C., Watson, B., & Winch, S. (2006). Extending a model of shift work tolerance. *Chronobiology International: The Journal of Biological and Medical Rhythm Research, 23*(6), 1363–1377. Retrieved November 9, 2007, from EBSCOHost database.

Price, W. H., Kiekbusch, R., & Theis, J. (2007). Causes of employee turnover in sheriff operated jails. *Public Personnel Management, 36*(1), 51–63. Retrieved April 6, 2007, from EBSCOHost database.

Ramlall, S. (2004). A review of employee motivation theories and their implications for employee retention within organizations. *Journal of American Academy of Business, Cambridge, 5*(1/2), 52–63. Retrieved September 30, 2006, from EBSCOHost database.

Results Based Interactions. (2004). *Leadership skills.* Development Dimensions International Inc., MCMXCIX, Pittsburgh, PA.

Riordan, C. M., Vandenberg, R. J., & Richardson, H. A. (2005). Employee involvement climate and organizational effectiveness. *Human Resource Management, 44*(4), 471–488. Retrieved April 25, 2007, from EBSCOHost database.

Rosen, A., & Callaly, T. (2005). Interdisciplinary teamwork and leadership: Issues for psychiatrists. *Australasian Psychiatry, 13*(3), 234–240. Retrieved November 24, 2007, from EBSCOHost database.

RoSPA. (2006). WSA winners announced. *Occupational Safety and Health Journal, 36*(4), 6. Retrieved July 26, 2008, from EBSCOHost database.

Schwieters, J., & Harper, D. (2007). Seven steps toward gaining control of your labor cost. *Healthcare Financial Management, 61*(4), 76–80. Retrieved August 27, 2007, from EBSCOHost database.

Sharbrough, W. C., Simmons, S. A., & Cantrill, D. A. (2006). Motivating language in industry: Its impact on job satisfaction and perceived supervisor effectiveness. *Journal of Business Communication, 43*, 322–343. Retrieved July 26, 2008, from EBSCOHost database.

Sharma, A., & Lutchman, C. (2006). Scope definition for expanding operating projects. American Association for Cost Engineers International Transactions. Retrieved March 29, 2008 from EBSCOHost database.

Shaw, J. D., Gupta, N., & Delery, J. E. (2005). Alternative conceptualizations of the relationship between voluntary turnover and organizational performance. *Academy of Management Journal, 48*(1), 50–68.

Silén-Lipponen, M., Tossavainen, K., Turunen, H., & Smith, A. (2004). Theatre nursing: Learning about teamwork in operating room clinical placement. *British Journal of Nursing, 13*(5), 244–297. Retrieved April 22, 2007, from EBSCOHost database.

Slattery, J., & Selvarajan, T. T. R. (2005). Antecedents to temporary employee's turnover intention. *Journal of Leadership and Organizational Studies, 12*(1), 53–66. Retrieved October 1, 2006, from EBSCOHost database.

Smither, L. (2003). Managing employee life cycles to improve labor retention. *Leadership and Management in Engineering, 3*(1), 19–23. Retrieved November 05, 2006, from EBSCOHost database.

Strom Clark, C. (2008). Women worth watching in 2009. *Profiles in Diversity Journal, 1010*(5), 59. Retrieved March 29, 2009, from EBSCOHost database.

Thompson, T. P. (2002). Turnover of licensed nurses in skilled nursing facilities. *Nursing Economics, 20*(2), 66–69. Retrieved April 11, 2007, from EBSCOHost database.

Thornton, S. L. (2001). How communication can aid retention. *Strategic Communication Management, 5*, 24–27. Retrieved November 3, 2006, from EBSCOHost database.

Tinham, B. (2008). Protect and perfect. *Works Management, 61*(9), 20–23. Retrieved January 10, 2009, from EBSCOHost database.

Trevor, C. O. (2001). Interactions among actual ease-of-movement determinants and job satisfaction in the prediction of voluntary turnover. *Academy of Management Journal, 44*(4), 621–638. Retrieved April 8, 2007, from EBSCOHost database.

Wilson, N., Cable, J. R., & Peel, M. J. (1990). Quit rates and the impact of participation, profit-sharing and unionization: Empirical evidence from UK engineering firms. *British Journal of Industrial Relations, 28*(2), 197–213. Retrieved April 25, 2007, from EBSCOHost database.

Wren, D. A. (1994). *The evolution of management thought* (4th ed.). New York: Wiley.

Zellars, K. L., Hochwarter, W. A., Perrewe, P. L., Miles, A. K., & Kiewitz, C. (2005). Beyond self-efficacy: Interactive effects of role conflict and perceived collective efficacy. *Journal of Managerial Issues, 13*(4), 483–500. Retrieved April 16, 2007, from EBSCOHost database.

Index